NEXUS NETWORK JOURNAL

ARCHES:
GATEWAYS FROM SCIENCE TO CULTURE

VOLUME 8, NUMBER 2

Autumn 2006

KIM WILLIAMS BOOKS

Nexus Network Journal
Vol. 8
No. 2
Pages 5–122
ISSN 1590-5896

CONTENTS

*O*n 19 February 2006, I directed the symposium entitled "Arches: Gateways from Science to Culture" at the AAAS Annual Meeting in St. Louis. The team of six speakers included Paul Calter (Vermont Technical College), Donald Hanlon (University of Wisconsin – Milwaukee), Santiago Huerta (Escuela Técnica Superior de Arquitectura de Madrid), Matthys Levy (Weidlinger Associates, NY), John Ochsendorf (MIT) and Michael Serra (author, *Discovering Geometry*).

The arch symposium team, from left, John Ochsendorf, Santiago Huerta, Michael Serra, Kim Williams, Paul Calter, Matthys Levy and Donald Hanlon

Taking its inspiration from St. Louis's Gateway Arch by Eero Saarinen and the theme of the AAAS meeting, "Grand Challenges, Great Opportunities," this symposium presented an exploration of the arch from the points of view of architecture, mathematics, engineering, construction history, and cultural symbolism. The arch, one of the most beautiful ways that architects invented to go from "here" to "there," spans greater distances and sustains larger loads than a simple post and beam structure, but because it is also more complex, an Eastern proverb called it "the structure that never sleeps." Leonardo da Vinci described the arch as "two weaknesses which, leaning on each other, become a strength," a metaphor for the way that science and art lean on each other to strengthen our lives.

I opened the symposium with an overview of the history and statistics of Eero Saarinen's Gateway Arch in the AAAS meeting's host city, St. Louis. At 630 feet, the arch is the tallest monument in the National Park system, a marvel of pre-computer science engineering and a fitting monument to Jefferson's dream of western expansion.

Renowned structural engineer Matthys Levy talk looked at historical development of the arch by the Egyptians, Greeks and Romans. Arch technology was known in Egypt and Greece, and arches are found, for instance, in service and underground structures in those cultures, but are not part of the vocabulary of their monumental architecture. The Romans, however, mastered arch construction on a large scale and made it the apex of their architectural expression, as in the triumphal arches.

MIT engineer John Ochsendorf explained that the characteristics of Saarinen's arch were well known to seventeenth-century English architect Robert Hooke, who explained that "[a]s hangs the flexible line, so stands the rigid arch." He then traced the development of the arch through the nineteenth-century work of Gustave Eiffel and the twentieth-century bridges of Robert Maillart. He discussed recent research at MIT on the interactive analysis of structural forms, which suggests new ways for arches to be used in design and construction in the future.

Santiago Huerta, president of Spain's society for construction history, discussed the historical geometrical rules for the proportional design of arches, independent of the scale, which were attacked by Galileo in the sixteenth century. In fact, Galileo's law applies only to strength problems, but not to problems of stability, which are indeed governed by geometry.

Mathematics professor and author Paul Calter examined the formula that governs the shape of the St. Louis arch, $y = 68.8(\cosh 0.01 - 1)$. Step-by-step, Prof. Calter broke the formula down to explain it in intuitive terms such as compound interest, the exponential growth of rabbit populations and the exponential decay of the temperature of a cup of coffee, then built it back up again to explain how it determines the shape of the arch.

Architect Donald Hanlon examined the symbolic function of the arch in terms of what it can express *for* – as well as *about* – a culture. The Roman triumphal arch, Paris's Arc de Triomphe and the Gateway Arch all testify to arch's symbolic capacity to embody and express the concepts of entry and passage. Referring to the earlier talk by Matthys Levy, Hanlon asked: if the Egyptians and Greeks were familiar with arch technology, why did they choose not to use it as a primary form? The answer lies in the relationship between those cultures and technology. We live in a culture that values technology in itself, so that all technological developments are immediately embraced. Ancient cultures, however, valued technology less than tradition, therefore although their architecture developed from wood to masonry construction, they continued to privilege the post-and-beam forms of wooden architecture.

Educator Michael Serra led symposium participants in a surprising and fun hands-on arch construction project using familiar objects—Chinese take-out cartons—in an unfamilar way: "these are stone voussoirs from an ancient miniature bridge uncovered by my friend, archaeologist Ertha Diggs. She has asked us to determine the number of stones in the original bridge." This makes it possible to understand both arch mechanics and the mathematics behind the arch through actually constructing them.

Other articles in this issue of the *Nexus Network Journal* also underline the special relationships between architecture and mathematics.

Elena Marchetti and Luisa Rossi Costa examine the mathematical rules that underlie the creation of surfaces in contemporary and historic architecture. Jose Iniguez and his student research group report on the Golden Mean and the Golden Quadratic Equation, $x^2 - x - 1 = 0$. Luisa and Victor Consiglieri present morphological rhythm in architecture. The issue is completed by Marcello Scalzo's review of a recent exhibit on Alberti in Florence, and book reviews by James McQuillan and myself. I hope you enjoy this special issue as much as I have.

Kim Williams

Matthys Levy

Weidlinger Associates Inc.
375 Hudson Street, 12th Floor
New York, NY 10014-3656 USA
levy@wai.com

keywords: arch construction,
Egyptian architecture, Greek
architecture, corbelled arches

Research

The Arch: Born in the Sewer, Raised to the Heavens

Abstract. The great ancient civilizations all knew about the arch yet the Greeks relegated its use to underground sewers and never raised an arch above ground: Why? The Egyptians, the Babylonians, the Assyrians and of course the Romans all exploited the arch as a means of spanning and enclosing space. Yet, curiously, Greece, one of the most cultured of the ancient civilizations and the builder of magnificent temples used stone in a most unnatural way, as a beam. The arch as a construction technique is intimately connected to the material of which it is constructed, namely masonry. Stone, when used as a beam is limited by size, scale and proportion. The Greeks certainly understood this as they closely spaced columns to support stone lintels. They also meekly tried to use stone in an A-frame configuration using a corbelled arch in Mycenea (1325 BC) but they never made the transition to the true arch using stone voussoirs. WHY? Perhaps the answer lies in a lack of understanding of the mechanics of the materials and the nature of compression and tension.

Early development of the arch

The great ancient civilizations all knew about the arch yet the Greeks relegated its use to underground sewers and never raised an arch above ground except in a minor or ornamental manner: Why? Examples abound in nature of arches such as the stone arch off Milos. Yet, curiously, Greece, one of the most cultured of the ancient civilizations and the builder of four of the seven wonders of the ancient world – the statue of Zeus at Olympia, the Temple of Artemis at Ephesus, the Colossus of Rhodes and the Lighthouse at Alexandria – and the builder of magnificent temples, used stone in a most unnatural way, as a beam. Even the architect of the Parthenon, Ictinos, thought that arches were not stable.[1]

The Egyptians, the Babylonians, the Assyrians and, of course, the Romans all exploited the arch as a means of spanning and enclosing space. The third-century B.C. Porta di Giove in the city wall of Falerii Novi (Etruria) and the first-century B.C. Pons Fabricus, a circular arch spanning less than 25m over the Tiber River in Rome, are typical examples. The Ponte Molle (Pons Mulvius) built by Marcus Aemilius Scaurus in 109 B.C. in Rome, with seven arched spans of 15-24m, is still standing.

The arch as a construction technique is intimately connected to the material of which it is constructed, namely masonry. Stone is limited by its size, scale and proportion when used as a beam. The Greeks certainly understood this, as they

1590-5896/06/020007-5 DOI 10.1007/s00004-006-0014-x

spaced columns closely to support stone lintels in temples such as the Parthenon. That structure used white pentelic marble from Mt Pentelicus in Attica, the same marble used for statues such as the Venus de Medici (sculpted by a Greek in the first-second century A.D.). One problem with marble is that, as a metamorphic rock derived from limestone, it is easily corroded by water and acid fumes. The disastrous deterioration of both statuary and temples witnessed in Athens is a direct result of this weakness.

Perhaps as a result of the observation of natural stone arches, Eupalinus of Megara led the excavation for a 1000m long water tunnel on the Greek island of Samos in 520 BC through a limestone mountain.[2] It was a magnificent engineering achievement. Eupalinus, the engineer for the venture, organized two crews working from either end. After working for ten years and without modern surveying apparatus, the tunnel deviated at the meeting point by less than 5m horizontally and 1m vertically. But for the purpose of our study, the interesting fact is the shape of the cross section that was about 1.8m in width and height with a curved soffit. Although not a true arch, it is clear that the arching concept of stone was understood by Eupalinus.

The Greeks also meekly tried to use stone using a (false) corbelled arch in the Lion Gate (1250 BC), an entrance to the Citadel, and in the Treasury of Atreus, the entrance to the Princes' tombs, both in Mycenea.[3] There are a number of other examples of stone arches used more in an ornamental fashion using stone voussoirs or true arches. The gateway to the Agora in Priene, built on a terrace above the harbor in 150 B.C., incorporated a 6m span arch. On the Acropolis, but below the Parthenon, arches can be seen. At Asclepion, built in 400 B.C. on the Greek island of Kos, off the Turkish coast where Hippocrates taught, a multiple-arched façade defines the Propylea, the baths and guest rooms of what was the oldest hospital in the world. South of Troy, the Gate of Asso is in the form of an arch and may actually predate the Gate at Volterra. Miletus had a similar arch. At Ephesus, the Library of Celcus has a number of decorative marble arches that date from a time after the classical Greek Era (700-400 B.C.). Further back in time, about 1360 B.C., the Hittite capital in Boghazkoy in Anatolia, east of Ankara, includes an elliptical arch as the King's Gate. Although not truly Greek, it was clearly an antecedent.

The masonry arch first appeared between the fifth and the fourth centuries B.C. in Greece, Etruria and Rome. Sometime in this period, the Etruscans most likely introduced the Romans to the arch. The Etruscan Gate at Volterra from the fourth century B.C. is considered the first example of a true arch. In Europe, the oldest known arch is the Cloaca Maxima, a huge drain built in 578 B.C. by Lucius Taquinius Priscus. The Romans adopted and developed the structural arch to an art form, such as at the forum of Thessalonica built in the second century A.D. along a Roman Road, but in Hellenistic architecture the arch was recognized as no more than a self-conscious tour de force.

What then were the reasons for the hesitant use of the arch by the Greeks in a more structural manner? One reason may have been a lack of understanding of the mechanics of the materials and the nature of compression and tension. Construction, in the days before the development of structural mechanics, always evolved as a trial-and-error process. It is quite likely that the arch as a structural element represented too much of a leap for the Greeks from simple post and lintel construction, perhaps because the thrust from an arch needs a lateral support and such support is easily provided below ground but needs special provision above ground.

Perhaps the material played a part! Consider the fact that the Romans built arches primarily of igneous granite or sedimentary limestones in such structures as the aqueduct in Segovia built during Emperor Trajan's time (ca. 100 A.D.) with 128 arches of white granite, and the 275m long Pont du Gard in Provence (14 A.D.), a three-tiered aqueduct 49m high constructed of limestone. They also used metamorphic rock such as tufa (a light volcanic igneous rock) and travertine marble in the Mulvian Bridge in Rome, which consists of 18m arches.

Strength of stone

Having learned from the Etruscans, the Romans acted confidently in their use of these different qualities of stone, from the strongest granites to the weaker marbles and limestones, and also took advantage of smaller stones which were adequate for an arch as compared to the larger stones needed for a lintel. The Greeks, on the other hand, built primarily of limestone or marble. Dense limestone and marble are both of about equal strength although less than half the strength of granite (see Table 1). However, limestone is a much more workable stone that can be cut or sawn more easily as well as splitting easily, having been deposited in layers. It is therefore well-suited for use in fluted columns and large lintels.

Stone	Compressive Strength	Flexural Strength
Granite	131 MPa	8.3 MPa
Marble	52 MPa	6.9 MPa
Limestone	12-55 MPa	n/a
Sandstone	28 MPa	n/a

Table 1.

Another possible reason against the construction of arches stems from the need for wood centering. In the Greek lands, the absence of forests made good quality timber a somewhat scarce commodity. Also, the skill needed to construct centering for the voussoirs may never have developed. This issue is not dissimilar to one faced in the recent past by designers wanting to build concrete shell structures. Such structures were very popular in the middle of the last century but are no longer being built. The reason is that centering and formwork is needed for this construction to define the shape and support the wet concrete. Rather than a question of scarcity of wood, the problem is the cost of the custom hand labor required to build the forms. For this reason shell structures are no longer being built, even though such structures are very efficient. It is really the same problem faced by the Greek builders wanting to erect arches that are in their own way much more efficient than the alternative post and lintel.

It is only in later years that the arch appeared in Greek construction such as the archway to the Kato Paphos in Cyprus built in the twelfth century and destroyed in the earthquake of 1222.

We may never know the true reason why structural arches did not appear in Hellenistic architecture but as argued above, the answer may be a combination of lack of structural experimentation, constraints presented by available stone and limitations on availability or cost of wood centering.

Notes

1. In *Math and the Mona Lisa* (Smithsonian Books, 2005), Bülent Atalay states: "Although the Greeks could make arches by fitting together wedge-shaped stones, they had not extended the idea into spanning large spaces by making domes of such stones."

2. Cited by Herodotus and the home of Agamemnon.

3. Examples of corbelled arches existed in many diverse civilizations such as the Maya and the Khmer.

About the author

Matthys P. Levy is a founding Principal and Chairman Emeritus of Weidlinger Associates, Consulting Engineers. Born in Switzerland and a graduate of the City College of New York, Mr. Levy received his MS and CE degrees from Columbia University. He has been an adjunct professor at Columbia University and a Distinguished Professor at Pratt Institute and a lecturer at universities throughout the world.

Mr. Levy is the recipient of many awards including the ASCE Innovation in Civil Engineering Award, the IASS Tsuboi Award, the ENR Medal of Excellence, three Lincoln Arc Welding awards, three PCI awards, the Founder's Award of the Salvadori Center and an AIA Institute Award. He was named a Structural Engineering Legend in Design by Structural Engineering Magazine in 2003. He has published numerous papers in the field of structures, computer analysis, aesthetics and building systems design, has illustrated two books and is the co-author of the best-selling book *Why Buildings Fall Dow* as well as *Structural Design in Architecture, Why the Earth Quakes, Earthquake Games and Engineering the City*. His latest book, *Adventures in Weather*, is looking for a publisher!

Mr. Levy is a member of the National Academy of Engineering, a Fellow of the Institution of Civil Engineers and the American Society of Civil Engineers, a member of the International Association of Shell & Spatial Structures, the International Association of Bridge and Structural Engineers and other professional societies. He is a registered Professional Engineer in the US and Eur Ing in Europe; he is also a founding director of the Salvadori Center that serves youngsters by teaching mathematics and science through motivating hands-on learning about the built environment.

Projects for which he was the principal designer include the Rose Center for Earth and Space at the American Museum of Natural History, the Javits Convention Center and the Marriott Marquis Hotel in New York, the Georgia Dome in Atlanta, the La Plata Stadium in Argentina, the One Financial Center tower in Boston, Banque Bruxelles Lambert in Belgium, the World Bank Headquarters in Washington, DC, and a cable-stayed pedestrian bridge at Rockefeller University. He is the inventor of the patented Tenstar Dome structure, a unique tensegrity cable dome used to cover large spaces with minimal obstruction.

Mr Levy was represented in the exhibit, 'The Engineer's Art' at the Centre Pompidou in Paris. He has appeared on numerous television shows including NOVA, Modern Marvels, the History Channel, ABC News, PBS series on Domes and others.

Mr. Levy has served as an expert in forensic investigations including the World Trade Center Collapses in New York, the Versailles Ballroom Collapse in Jerusalem, the failure of the UNI Dome in Iowa, the Fire Damage to the Meridian Building in Philadelphia, the collapse of Precast Concrete Stands in Atlanta, the quality of Aquarium Construction in Duluth, the adequacy of documents for construction of a Multi-Screen Movie theatre in San Francisco, the cause of fabric roof failure in Montreal.

Philippe Block
Matt DeJong
John Ochsendorf

Building Technology Program
MIT Room 5-418
Cambridge MA 02139 USA
ph_block@mit.edu
mdejong@mit.edu
jao@mit.edu

Keywords: limit analysis,
graphic statics, thrust line,
masonry arch, funicular,
earthquake assessment,
masonry

Research

As Hangs the Flexible Line:
Equilibrium of Masonry Arches

Abstract. In 1675, English scientist Robert Hooke discovered "the true... ...manner of arches for building," which he summarized with a single phrase: "As hangs the flexible line, so but inverted will stand the rigid arch." In the centuries that followed, Hooke's simple idea has been used to understand and design numerous important works. Recent research at MIT on the interactive analysis of structural forces provides new graphical tools for the understanding of arch behavior, which are useful for relating the forces and geometry of masonry structures. The key mathematical principle is the use of graphical analysis to determine possible equilibrium states.

Introduction

Robert Hooke's hanging chain. Robert Hooke (1635-1703) described the relationship between a hanging chain, which forms a catenary in tension under its own weight, and an arch, which stands in compression (fig. 1a).

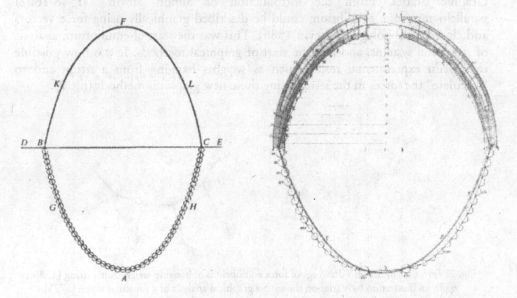

Fig. 1. (a) Poleni's drawing of Hooke's analogy between an arch and a hanging chain, and (b) his analysis of the Dome of St.-Peter's in Rome [1748]

Though he could not derive the equation of a catenary, Hooke knew that his intuition was right and therefore wrote his finding as an anagram in Latin in the margin of another book [Hooke 1675]. Once descrambled, the anagram reads: *ut pendet continuum flexile, sic stabit contiguum rigidum inversum*, and translates to "*as hangs the flexible line, so but inverted will stand the rigid arch*" [Heyman 1998]. Both the hanging chain and the arch must be in equilibrium, and the forces are simply reversed. The chain can support only tension, and the masonry arch acts in compression. Generalized, this idea signifies that the shape a string takes under a set of loads, if rigidified and inverted, illustrates a path of compressive forces for an arched structure to support the same set of loads. This shape of the string and the inverted arch is called a *funicular* shape for these loads.

In 1748, Poleni analyzed a real structure using Hooke's idea to assess the safety of the cracked dome of St. Peter's in Rome. Poleni showed that the dome was safe by employing the hanging chain principle. For this, he divided the dome in slices and hung 32 unequal weights proportional to the weight of corresponding sections of that "arch" wedge, and then showed that the hanging chain could fit within the section of the arch (fig. 1b). If a line of force can be found that lies everywhere within the masonry, then the structure can be shown to be safe for that set of loads [Heyman 1966].

Graphic Statics. From the introduction of Simon Stevin's (1548-1602) parallelogram rule, equilibrium could be described graphically using force vectors and closed force polygons [Stevin 1586]. This was the start of equilibrium analysis of structural systems, and also the start of graphical methods. It was now possible to explain experimental results such as weights hanging from a string and to "calculate" the forces in the string using these new graphical methods (fig. 2).

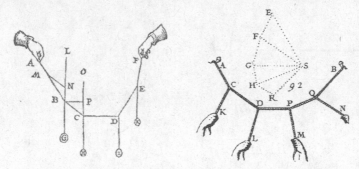

Fig. 2. *Left*, One of Stevin's drawings of force equilibrium of hanging weights on a string [1586]; *right*, an illustration by Varignon showing a graphical analysis of a funicular shape [1725]

Culmann [1866] was the first to formalize graphical analysis as a powerful method for equilibrium analysis in structural engineering. His *Die graphische Statik* had a strong theoretical foundation in mathematics, specifically in projected geometry. Graphical analysis provides a rigorous analysis method for trusses, arches, cables, and other structural systems. At the end of the nineteenth and the beginning of the twentieth century, graphic statics was the most common method to determine equilibrium for structures. Many trusses and masonry arch bridges were calculated using graphical methods, which still stand without any significant failures [Lévy 1888]. Handbooks from the beginning of the twentieth century, such as Wolfe [1921], use graphical constructions to solve advanced structural problems that would demand higher order differential equations when solved numerically. Maurer [1998] provides an excellent historical overview of the development and evolution of Culmann's method.

Even though graphic statics was successfully used in engineering practice, its popularity did not last long. By 1920, graphical methods were largely replaced by the theory of elasticity, which provided elegant closed-form analytical solutions that did not require the same drawing skills. As Boothby [2001] pointed out, graphical methods give good but conservative results, though the process and analysis can become very tedious. Few engineers or architects today have the specific knowledge and patience to make these complex and advanced graphic constructions. In recent years, Zalewski and Allen [1998] have published a new textbook extolling the virtues of graphic statics for design, which suggests that there are new applications for this historical analysis method.

Analysis of Masonry Arches. There are only three types of equations that can be used for structural analysis: equilibrium (statics); geometrical (compatibility), and materials (stresses). For historical masonry structures, the first two types of equations are most important, since stresses are typically an order of magnitude below the failure stress of the masonry. A stability or equilibrium approach will therefore be most valuable to assess the safety of masonry structures, and limit analysis provides a theoretical framework. To apply limit analysis to masonry, Heyman [1966] demonstrated that it is necessary to make three main assumptions: masonry has no tensile strength; it can resist infinite compression; and no sliding will occur within the masonry. Heyman [1982, 1995] and Huerta [2004] provide additional background on limit analysis for masonry arches.

Equilibrium in a masonry arch can be visualized using a *line of thrust*. This is a theoretical line that represents the path of the resultants of the compressive forces through the stone structure. This line is the inverted catenary discussed above. For a pure compression structure to be in equilibrium with the applied loads there must be a line of thrust that lies entirely within the masonry section. The concept was first rigorously formulated by Moseley [1833] and an excellent mathematical treatment was offered by Milankovitch [1907]. The analytical solution for this concept has been defined more precisely as the *locus of pressure points* by Ochsendorf [2002].

New interactive equilibrium tools

Recent research at MIT has produced new computer methods for graphic statics. Greenwold and Allen [2003] developed *Active Statics,* a series of interactive online tutorials that implement graphic statics for a series of chosen problems. As a continuation, new methods for exploring the equilibrium and compatibility of masonry structures have been developed by Block et al. [2005] which can be applied in real-time. In the tradition of limit analysis developed by Heyman, thrust line analysis is used to provide a clear means for understanding the behavior and safety of traditional masonry structures. Limit analysis using thrust lines can establish the relative stability of the structures as well as possible collapse mechanisms.

This paper presents a series of models and tools demonstrating the possibilities of this approach for analysis, design and teaching. This new approach brings back graphical analysis, greatly enhanced by the use of computers, which allow for parametric models of structural elements or systems. Their geometry is linked to the graphical construction and controls the loads of the analysis. A rigid block model is used to provide displacement or kinematic analysis and animations illustrate collapse. The models are interactive and parametric, allowing the user to change all parameters (such as thickness, height, span etc.) in order to explore an entire family of structural shapes and to understand the relation between geometrical changes and stability. At the same time, the methods are numerically accurate and rigorous.

The three new ideas of this approach are: interactive graphic statics, geometry controlled loads, and animated kinematics. Block et al. [2005] provides a detailed explanation of these steps.

1. Interactive graphic statics:
   The models are made using simple two-dimensional drawing packages, such as Cabri Geometry [2006]. They allow the creation of dynamic geometric constructions, and to make these easily available over the Internet. Computer-aided drawing programs overcome the inconveniences and drawbacks of graphic statics by guaranteeing accuracy. Furthermore, parametric modeling prevents the user from having to redo the graphical constructions for every analysis.

2. Loads controlled by the geometry:
   The geometry of the structural element is linked to the graphical construction. Altering the section on the screen will influence the self-weight and this is updated instantly thereby changing the applied loads. This results in an interactive tool that provides real-time structural feedback.

3. Animated kinematics:
   The assumptions necessary to apply limit analysis to masonry structures allow displacement analysis to be performed in real time. The blocks in the models stay rigid during the kinematic movement and their movement can be described using geometry.

Examples

This section presents examples of tools made possible with this new approach. The project website at *http//web.mit.edu/masonry/interactiveThrust* contains more in-depth analysis.

Thrust-line analysis of a random arch. Fig. 3 uses (a) Bow's notation and (b) a force polygon to give the magnitude of the forces of the segments in the funicular polygon for a random arch. This force polygon is drawn to its own scale and represents and visualizes the equilibrium of the system. The funicular construction and its polygon are related by geometry. The mathematical foundations for this reciprocal relationship are clearly summarized by Scholtz [1989]. The horizontal distance from the pole *o* to the load line *ah* gives the horizontal thrust in the system. This is the amount the arch pushes (thrusts) outwards and analogously the amount the hanging string pulls inwards. Looking more closely at the fan shape of the force polygon, we can isolate the different closed vector triangles (bold lines) showing the equilibrium of each block in the random arch (Fig. 3c, d).

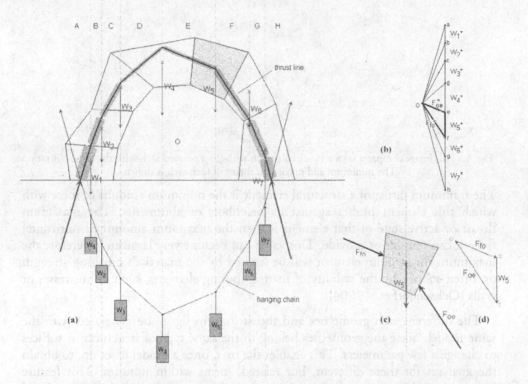

Fig. 3. For a random arched structure, (a) a possible thrust line and its equivalent hanging chain are constructed using graphic statics; (b) the force equilibrium of the system is represented in the funicular polygon; (c) the equilibrium of one of the *voussoirs*; and, (d) the vectors representing the forces in and on the block

First, the structure has to be divided into discrete parts. The actions of the different blocks are treated as lumped masses applied at their center of gravity (Fig. 3a). The magnitudes of the forces are proportional to their weight and transferred to the force polygon (b). If the user drags a corner point, this would influence the area, hence the weight, of the two adjacent blocks. This will then also alter the associated vector in the force polygon.

It can also be seen that there is not only one solution to this problem since a different horizontal thrust in the system results in a deeper (for decreased thrust) or a more shallow (for increased thrust) section. Thanks to the interactive setup the user can explore this solution easily by controlling the unknown in the "equations", for example the amount of horizontal thrust in the system. The notion that there is no one answer to this problem will be expanded in the following section.

Semi-circular versus pointed arch. For every masonry structural element, there is an infinite amount of valid thrust lines that fit within its section, all lying between a maximum and minimum value (fig.4).

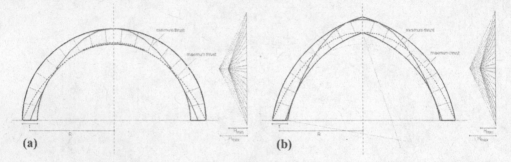

(a)   (b)

Fig. 4. This image compares (a) a semi-circular arch with (b) a pointed arch with the same t/R ratio. The minimum and maximum thrust of each arch is shown

The minimum thrust of a structural element is the minimum amount of force with which this element pushes against its neighbors or abutments. The maximum thrust or active state of that element is then the maximum amount of horizontal force it can transfer or provide. This value can become very large and therefore the maximum thrust of an element will be limited by the material's crushing strength or, often sooner, by the stability of its neighboring elements, such as buttresses or walls [Ochsendorf et al 2004].

The different arch geometries and thrust lines in fig. 4 are generated with the same model. Since the geometries belong to the same type of structures, it suffices to change a few parameters. This enables the user, once a model is set up, to obtain the analyses for these different, but related, forms within minutes. This feature becomes very interesting when a vast number of structures have to be analyzed and compared, as applied to flying buttresses by Nikolinakou et al [2005].

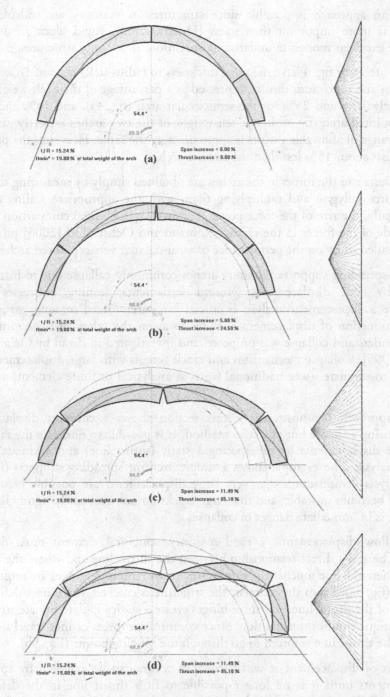

Fig. 5. This shows an arch on spreading supports with a t/R = 15% and a total angle of embrace of 160° (a) initially in its minimum thrust state, (b) at an inter-mediate state, and (c) right at collapse. (d) shows a snapshot of the animation showing the collapse mechanism

Such an approach is possible since structures in masonry are scalable, and stability is more important than stress [Huerta 2004]. Rigid block models are therefore excellent models to understand unreinforced masonry structures.

The two arches of fig. 4 have the same thickness to radius (t/R) ratio of 18%. Their minimum and maximum thrusts, expressed as a percentage of their self-weight, are respectively 16% and 25% for the semicircular arch (fig. 4a), and 14% and 23% for the pointed arch (fig. 4b). The self-weight of the two arches is nearly identical and the range of allowable thrusts is approximately the same. However, the pointed arch thrusts about 15% less than the circular arch.

These results and the forces in the arches are obtained simply by measuring the rays of the force polygon and multiplying them with the appropriate scaling factor. Additionally, the size of the force polygon allows a quick visual comparison of the magnitude of the forces in the system. Romano and Ochsendorf [2006] provide a more detailed study on the performance of semi-circular versus pointed arches.

Arch on spreading supports. Masonry arches commonly collapse due to instability caused by large displacements (ground settlements, leaning buttresses etc.). Therefore a displacement analysis is crucial for unreinforced masonry structures. The introduction of displacement analyses to assess the safety of these structures and to understand collapse was proposed and investigated in detail by Ochsendorf [2002, 2005]. Collapse mechanisms and crack genesis with large displacements are not easy to simulate using traditional (such as analytical or finite element) analysis methods.

The approach demonstrated in this section shows a complex displacement analysis using a simple but powerful method. It is possible to illustrate the range of allowable displacements by superposing a static (thrust line) and a kinetic (rigid body) analysis. The example shows a simple arch on spreading supports (fig. 5). The analysis demonstrates that very large displacements are possible before the structure becomes unstable, and that large cracks do not necessarily signify that the structure is in immediate danger of collapse.

To allow displacements, a rigid masonry structural element must develop cracks. There is a direct relationship between the thrust line and where the hinges occur: where the line touches the extremities of the structure, hinges are most likely to form (fig. 5a). From the moment the structure is cracked, the hinges define the location of the thrust line, and three hinges create a statically determinate structure with a unique equilibrium solution. Since compressive forces cannot travel through the air, the thrust line is forced to go through the hinging points (Fig. 5b).

The limit of displacement at which collapse occurs can be explored by applying displacements until it is no longer possible to fit a thrust line in the deformed structure. Failure occurs when there are more than three hinges, i.e., when the thrust line touches the boundaries of the structure in more than three places (fig.

5c). Fig. 5d shows a snapshot of an animation of a possible collapse mode due to spreading supports.

This model is interactive; the effects of spreading supports can be checked by the user by moving the corner or inputting values. While doing this, it can be noticed that the force polygon grows, indicating that the forces, specifically the horizontal thrust, in the system increase. This makes sense since the deforming arch becomes more and more shallow, resulting in an increase of horizontal thrust.

Earthquake Analysis of Masonry Arches. This approach can also be extended to assess the stability and safety of arched structures in areas prone to earthquakes. However, the assumptions adopted from Heyman [1995] must be reevaluated for dynamic loading. Clearly dynamic loading could increase local stresses causing crushing failure of the masonry, and could cause vibrations which would increase the likelihood of sliding. Despite the fact that such a purely geometrical analysis method cannot capture these more complicated effects of earthquake dynamics, the interactive graphical analysis can be useful.

It is common practice in structural engineering design to simulate earthquake loading by a constant horizontal force that is some fraction of the weight of the structure in magnitude. This is equivalent to applying a constant horizontal acceleration that is some fraction of the acceleration of gravity. Such an "equivalent static loading" does not capture all of the dynamics, but it does provide a measure of the lateral loading that the structure could withstand before collapse.

To implement equivalent static analysis for earthquake loading simulation, it is possible to modify the interactive thrust tool to include a tilting ground surface. Tilting the ground surface effectively applies a vertical acceleration (gravity, g) and a horizontal acceleration (λ*g, where $\lambda = \tan(\alpha)$ and α is the ground surface inclination angle). The ground surface is tilted until the thrust line cannot be contained within the structure: the point where it passes through the exterior surface of the structure at four locations. At this point, four 'hinges' would form and the structure would collapse. Two examples of structures subjected to equivalent static analysis are shown at the point of collapse in Fig. 6.

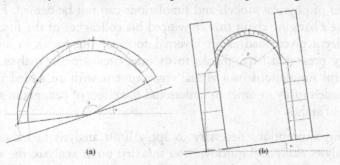

Fig. 6. Equivalent static analysis of (a) a masonry arch, and (b) a similar arch on buttresses. Points indicate hinge locations at collapse

The arch of sixteen voussoirs, inclusion angle (β) of 157.5 degrees, and thickness/radius ratio of 0.15 would collapse at a constant horizontal acceleration of 0.34^*g (fig. 6a). A similar arch (t/r = 0.15, b = 120°) on buttresses with relative height and width dimensions of 1.5 and 0.5 times the vault span, respectively, would collapse at a constant horizontal acceleration of 0.13^*g (fig. 6b). These equilibrium solutions can be verified through other common analysis techniques such as equilibrium equations or virtual work.

The resulting values of horizontal accelerations which cause collapse are conservative in that they assume an infinite duration of loading. Actual earthquake loadings are of much shorter duration which could allow the structure to "recover" due to inertial effects if higher horizontal accelerations are experienced. However, the analysis is also unconservative because the possibilities of local crushing and sliding have not been evaluated. Regardless, the method is valuable because it provides a rapid relative measure of the stability of vaulted masonry structures, and visually depicts the expected collapse mechanism. Therefore, while more rigorous dynamic analyses should be executed for structures which are known to be in danger of collapse, this equivalent static analysis method can provide a valuable tool for identifying those structures.

Conclusion

As has been shown through previous examples, the interactive analysis tools provide a useful method which uses graphic statics to achieve a rapid first order assessment of the stability of various masonry arch structures. Specifically, the real time graphic statics framework has been shown to allow the effects of geometrical changes such as arch thickness, buttress width, etc., to be quickly evaluated. Assessment of the stability of such arched structures with varying geometries would be considerably more difficult using traditional computational methods, but more importantly, the graphic statics framework inherently presents results visually as well as numerically, allowing results to be easily interpreted. Mistakes cannot hide in the equations and the validity of the results is guaranteed because of the visual nature of the technique.

The power of physical models and simulations can not be denied. For example, it was Hooke's hanging chain that convinced his colleagues of the forces acting in tension under a given load as an inverted form of the forces in an arch. The methodology presented here, thanks to its integrated kinetic analysis, allows the same powerful simulations in a virtual environment with an added versatility to adapt the models easily in order to understand the effect of changes in geometry on the stability of arches.

The strong assumptions necessary to apply limit analysis to masonry provide complex analyses easily and quickly. After this first order analysis, the analyst must check if the assumptions have been violated. In some cases, the method will lead to unconservative results since second order effects such as local crushing, crack propagation and sliding have not been considered. This is most crucial in the kinetic analyses.

This paper has shown that there is great potential in using interactive thrust line analysis for masonry structures. It has also raised new research questions to be developed and has given possible paths to be considered for further exploration and development. The paper showed how graphical computation offers new possibilities in an old field of research, which is essential for conserving architectural heritage in the future. Finally, the concept of the thrust line emphasizes the relationship between geometry and structural behavior of buildings as a fundamental principle for architectural designers in the future.

References

BLOCK, P. 2005. *Equilibrium Systems. Studies in masonry structure.* M.S. dissertation, Department of Architecture, Massachusetts Institute of Technology, June 2005.

BLOCK, P., T. CIBLAC, and J.A. OCHSENDORF. 2005. Real-Time Limit Analysis of Vaulted Masonry Buildings. *Computers and Structures* (Submitted for review July 2005).

BOOTHBY, T.E. 2001. Analysis of masonry arches and vaults. *Progress in Structural Engineering and Materials* 3: 246-256.

CABRI GEOMETRY II PLUS. 2006. Cabrilog sas, France. http://www.cabri.com/

CULMANN, K. 1866. *Die graphische Statik.* Zürich : Meyer und Zeller.

DEJONG, M. and J.A. OCHSENDORF. 2006. Analysis of vaulted masonry structures subjected to horizontal ground motion. In *Structural Analysis of Historical Constructions,* P.B. Lourenço, P. Roca, C. Modena, S. Agrawal, eds. New-Dehli.

GREENWOLD, S. and E. ALLEN. 2003. *Active Statics.* Cambridge, MIT. http://acg.media.mit.edu/people/simong/statics/data/index.html

HEYMAN, J. 1966. The stone skeleton. *International Journal of Solids and Structures,* 2: 249-279.

———. 1982. *The Masonry Arch.* Chichester: Ellis Horwood.

———. 1995. *The Stone Skeleton: Structural engineering of masonry architecture.* Cambridge: Cambridge University Press.

———. 1998. *Structural Analysis: A Historical Approach.* Cambridge: Cambridge University Press.

HOOKE, R. 1675. *A description of helioscopes, and some other instruments.* London.

HUERTA, S. 2004. *Arcos bóvedas y cúpulas. Geometría y equilibrio en el cálculo tradicional de estructuras de fábrica.* Madrid : Instituto Juan de Herrero.

———. 2005. The use of simple models in the teaching of the essentials of masonry arch behaviour. Pp. 747-761 in *Theory and Practice of Construction: Knowledge, Means, and Models,* Ravenna: Ed. G. Mochi.

LEVY, M. 1888. *La Statique Graphique et ses Applications aux Constructions.* Paris: Gauthier-Villars.

MAURER, B. 1998. *Karl Culmann und die graphische Statik.* Berlin/Diepholtz/Stuttgart: Verlag für die Geschichte der Naturwissenschaft und der Technik.

MILANKOVITCH, M. 1907. Theorie der Druckkurven. *Zeitschrift für Mathematik und Physik* 55: 1-27.

MOSELEY, H. 1833. On a new principle in statics, called the principle of least pressure. *Philosophical Magazine* 3: 285-288.

NIKOLINAKOU, M., A. TALLON and J.A. OCHSENDORF. 2005. Structure and Form of Early Flying Buttresses. *Revue Européenne de Génie Civil* 9, 9-10: 1191-1217.

OCHSENDORF, J.A. 2002. Collapse of masonry structures. Ph.D. dissertation, Department of Engineering, Cambridge University, June 2002.

———. 2006. The Masonry Arch on Spreading Supports. *The Structural Engineer,* Institution of Structural Engineers. London, Vol. 84, 2: 29-36.

OCHSENDORF, J.A., S. HUERTA and J.I. HERNANDO. 2004. Collapse of masonry buttresses. *Journal of Architectural Engineering*, ASCE 10 (2004): 88-97.

POLENI, G. 1748. *Memorie istoriche della Gran Cupola del Tempio Vaticano*. Padua: Nella Stamperia del Seminario.

ROMANO, A. and J.A. OCHSENDORF. 2006. Masonry circular, pointed and basket handle arches: a comparison of the structural behavior. In *Structural Analysis of Historical Constructions*, P.B. Lourenço, P. Roca, C. Modena, S. Agrawal (Eds.) New-Dehli.

SCHOLTZ, E. 1989. *Symmetrie – Gruppe – Dualität: Zur Beziehung zwischen theoretischer Mathematik und Anwendungen in Kristallographie und Baustatik des 19. Jahrhunderts*. Basel : Birkhäuser Verlag.

STEVIN, S. 1586. De Beghinselen der Weeghconst. (In *The Principal Works of Simon Stevin*, vol. 1, Leyden, 1955).

VARIGNON, P. 1725. *Nouvelle mécanique ou statique*. 2 vols. Paris.

ZALEWSKI, W. and E. ALLEN. 1998. *Shaping Structures*. New York: John Wiley & Sons.

About the authors

Philippe Block is a PhD candidate and Research Assistant in Building Technology at MIT. He studied architecture and structural engineering at the Vrije Universiteit Brussel in Belgium before coming to MIT for a master's thesis in Building Technology. His thesis developed interactive and parametric analysis tools for masonry structures.

Matt DeJong is a PhD candidate and Research Assistant in Building Technology at MIT. He completed his undergraduate studies in civil engineering at the University of California, Davis before working for three years as a structural engineer. He received his S.M. in Structures and Materials at MIT, and his current PhD research involves the analysis of unreinforced masonry structures subjected to earthquake loading.

John Ochsendorf is a structural engineer and Assistant Professor of Building Technology at MIT. He studied civil engineering at Cornell University and Princeton University before earning a PhD in structural mechanics at Cambridge University. He worked for one year as a Fulbright Scholar at the Universidad Politécnica de Madrid and now leads a research group on masonry mechanics at MIT [http://web.mit.edu/masonry].

Santiago Huerta

E. T.S. de Arquitectura
Universidad Politécnica de
Madrid
Avda. Juan de Herrera, 4
28040 Madrid SPAIN
Santiago.Huerta@upm.es

Keywords: arch design,
masonry arches, history of
engineering, history of
construction, structural
design, Galileo, strength of
materials

Research

Galileo was Wrong: The Geometrical Design of Masonry Arches

Abstract. Since antiquity master builders have always used simple geometrical rules for designing arches. Typically, for a certain form, the thickness is a fraction of the span. This is a proportional design independent of the scale: the same ratio thickness/span applies for spans of 10m or 100m. Rules of the same kind were also used for more complex problems, such as the design of a buttress for a cross-vault. Galileo attacked this kind of proportional design in his *Dialogues*. He stated the so-called square-cube law: internal stresses grow linearly with scale and therefore the elements of the structures must become thicker in proportion. This law has been accepted many times uncritically by historians of engineering, who have considered the traditional geometrical design as unscientific and incorrect. In fact, Galileo's law applies only to strength problems. Stability problems, such as the masonry arch problem, are governed by geometry. Therefore, Galileo was wrong in applying his reasoning to masonry buildings

Introduction

Arches are the essential element of masonry construction. They were invented some 6,000 years ago in Mesopotamia. The first arches were small; they were used to cover tombs. It is fascinating to look at the structural experimentation which developed over the course of 2,000 years before the arches emerged from the earth and began to form part of architecture proper [Besenval 1984; El-Naggar 1999]. The barns of the Ramesseum (thirteenth century BC) were covered by barrel vaults and the Hanging Gardens of Babylon (seventh century BC) where, in fact, supported on a system of arches and vaults. In Europe the Etruscans where among the first to use stone arches of moderate size, but it was in Imperial Rome, when the arch and the vault began to form an essential part of architecture, that the spans grew in order of magnitude, reaching the 43m of the Pantheon. Since then and up to the nineteenth century arches and vaults were at the heart of structural design: "All the aspects of architecture are derived from the vault", said Auguste Choisy, and the history of the different architectonic styles is also the history of how to solve the technical problem of building arches and vaults of different forms in brick, stone and mortar.

An arch thrusts: the stones, trying to fall down due to the force of gravity, produce inclined forces which are transmitted within the masonry down the springings (fig. 1). The forces must be inclined to give vertical equilibrium, but at the same time they produce a horizontal component. The vertical loads increase from top (the keystone) to the springings, but the horizontal component remains constant all through the arch to maintain horizontal equilibrium. At the springings of the arch there is, then, an inclined force which must be resisted by the *buttresses*. The structural design of masonry architecture deals with two fundamental problems: 1) To design arches which will stand; 2) To build buttresses which will withstand their thrust. Roman solutions are very different from Gothic or Baroque solutions, but the problem remains the same: to obtain a safe state of equilibrium which will guarantee the life of the building for centuries or millennia.

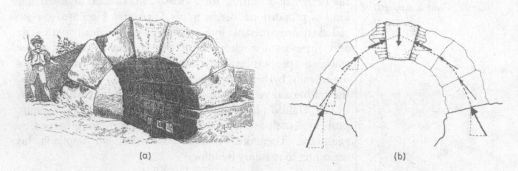

(a)   (b)

Fig. 1. (a) Etruscan voussoir arch [Durm 1885]; (b) Equilibrium of the stones in an arch. The inclined forces are transmitted within the arch and at the abutments there is always a thrust with a horizontal component (uniform through the entire arch) which must be resisted

The question is, how were arches and buttresses designed? The scientific theory of structures was applied only during the nineteenth century and this fact leaves almost all historical architecture outside the realm of this branch of modern applied mechanics. However, it is evident that the great buildings of the past could not have been built without some kind of knowledge: the master builders used a theory of a different kind, based in the critical observation of masonry building processes. This "non-scientific" theory must have been rich and complex, because its application resulted in the Pantheon, the Gothic cathedrals and the Hagia Sophia.

Proportional design of arches, vaults and buttresses. We know that this traditional theory, or the particular expression of it in every particular epoch, was condensed in the form of structural rules. For example, to design an arch of a certain form, the relevant parameter is the thickness, and this was always obtained as a fraction of the span. The same occurs with the buttresses, whose depth was calculated, again, as fraction of the span. Of course, the rules were specific for each structural type: the Gothic rules for buttresses give the depth of the buttresses as nearly 1/4 of the span; Renaissance rules give between 1/3 and 1/2 of the span. Gothic vaults are much lighter than Renaissance vaults, but the approach is the same.

The beginning of one of the manuscripts which have survived from the late Gothic period (*Vom des Chores Mass* (The measure of the Choir) published in [Coenen 1990]) expresses the method in a clear way:

> The building follows precise laws and all its parts are ruled, in such a way that all its elements are related with the whole building and the whole building is related with each one of its parts. The Choir is the fundament and the origin of all the rules, and from its span we obtain not only the thickness of its wall, but also the templates of the imposts and of all the elements of the work.

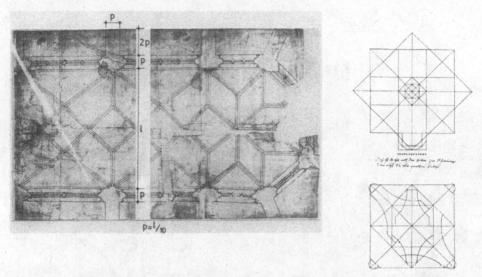

Fig. 2. Late Gothic structural design of a church and its elements. *Left,* the span is the "great module" from which all the elements are derived (letters superposed on an original Gothic drawing from [Koepf 1969]). *Right,* the wall thickness (span/10) is the "little module" from which the ribs, mullions and imposts are obtained, employing the *ad quadratum* technique of rotating squares [Hecht 1979].

The span of the vault of the choir, let us call it s, is a "great module" and all the dimensions of the structural elements are obtained as fractions of it. The wall should be $s/10$ and the buttresses three times this quantity, i.e., $3s/10$. The templates of the ribs and the mullions of the windows were also obtained from the wall thickness (fig. 2).

There were other rules for buttress design as well which have survived through the successive copies and re-elaborations of the medieval stone-cutting manuals, part of which were incorporated in Renaissance and Baroque stone-cutting handbooks. In fig. 3 a geometrical rule is represented. The intrados of the transverse arch of the vault is divided in three parts and a line is traced joining one of the points with the springings. Then the same distance is taken as the prolongation of the line and this point gives the depth of the buttress.

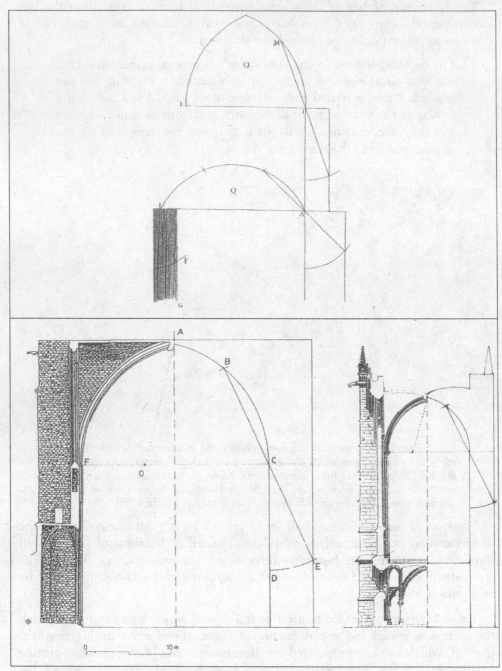

Fig. 3. Gothic geometrical rule for buttress design. *Above*, explanation of the rule [Derand 1643]. *Below*, application of the rule to two different buildings drawn to the same scale: the cathedral of Gerona and the Sainte Chapelle of Paris [Huerta 2004].

Rules of the same kind, arithmetical or geometrical, were employed in other periods as well – the Renaissance, Baroque, etc. – until in the nineteenth century masonry architecture began an accelerated decay, due to the appearance of new materials (iron, steel, reinforced concrete) and new structural types (frames, trusses, thin shells). However, structural design rules were also used all through the nineteenth century.

The essential characteristic of all these rules is that they are "proportional" and that they control the overall form of the structure of the building. It is a "geometrical design", which was considered to be correct for a building of any size.

Galileo and the "square-cube law"

Galileo (1564–1642) was the first to treat structural problems in a scientific way. During his forced reclusion in Arcetri, he wrote a book entitled *Discorsi e Dimostrazioni Matematiche intorno à due nuove sicenze Attenenti alla Mecanica & i movimenti Locali* (*Dialogues Concerning Two New Sciences*), published in 1638 (fig. 4). The two sciences were the Strength of Materials and the Cinematics, two topics less problematic than cosmology. It is the strength of materials which interests us.

Fig. 4. *Left,* portrait of Galileo Galilei. *Right,* first page of the *Dialogues (Discorsi)* of 1638

Galileo was trying, for the first time, to draw scientific conclusions about the strength of beams, a problem of evident practical interest. However, just from the beginning he exposes the result of his researches and mounts an attack on medieval proportional design (this and the following quotations are from the 1954 translation of Crew and de Salvio):

> Therefore, Sagredo, you would do well to change the opinion which you, and perhaps also many other students of mechanics, have entertained concerning the ability of machines and structures to resist external disturbances, thinking that when they are built of the same material and maintain the same ratio between parts, they are able equally, or rather proportionally, to resist or yield to such external disturbances and blows. For we can demonstrate by geometry that the large machine is not proportionally stronger that the small. Finally we may say that, for every machine and structure, whether artificial or natural, there is set a necessary limit beyond which neither art nor nature can pass; it is here understood, of course, that the material is the same and the proportion preserved.

The strength of beams. Galileo, however, considers only the case of simple bending. It is obvious that a column of any material will possess an absolute strength, the force necessary to break the column in tension. This force is proportional to the area of the cross-section (fig. 5, left). Now, Galileo is interested in obtaining the strength of a beam for which the absolute strength is known. Galileo chooses the simple case of a cantilever beam.

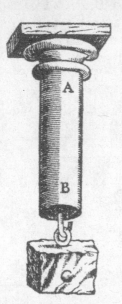

Fig. 5. *Left*, hypothetical trial to ascertain the absolute strength of a column. *Right*, the problem of the cantilever: How much weight could the cantilever sustain? [Galileo 1638]

He considers that the moment of resistance will be the result of multiplying the absolute strength by half the depth of the beam, as if a "hinge" (fulcrum) will form in the lower part of the critical section, where the beam is embedded into the wall. The analysis of Galileo is incorrect, as he forgets the necessary horizontal force to establish the equilibrium of horizontal forces: a member in bending must have both tension and compression areas. The correct solution was hinted at fifty years after by Mariotte in 1686; the first correct analysis was made by Parent in 1713; in 1773 Coulomb presented a complete description of simple bending theory. However, Galileo's error persisted in some handbooks until after 1800. The problem has been fully discussed in [Heyman 1998].

The square-cube law. However, Galileo was right about the form of the equation: for a given material and a certain cross-section, the bending strength is proportional to the product of its area by its depth. Galileo, then, applies himself to a comparison of the strength of beams of the same material and section but of different sizes. If the only load is the self-weight of the beam, Galileo realized that this load will grow with the third power of the linear dimensions, if the beam scales up maintaining its geometrical form. However, the strength will grow with the second power, the square, of the linear dimensions. As a consequence a structure becomes "weaker" as it grows in size, the reserve of strength diminishing linearly with size. If one wants to maintain the same strength, then, the cross-section must become thicker.

Galileo realized that this argument was a big discovery and immediately explained the consequences: "From what has already been demonstrated, you can plainly see the impossibility of increasing the size of structures to vast dimensions either in art or in nature". Therefore it will be impossible to build "ships, palaces, or temples of enormous size". Also, the size of an animal cannot be increased "for this increase in height can be accomplished only by employing a material which is harder and stronger than usual or by enlarging the size of the bones, thus changing their shape until the form and appearance of the animals suggest a monstrosity". Galileo then compares the deformation which will produce an increase of size of three times in the bone of an animal (fig. 6). The drawing explains the argument much better than the discussion in the text, and has been reproduced hundreds of times in texts about structures or biology.

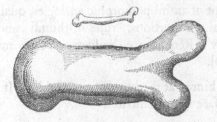

Fig. 6. The effect of an increase of size in the bones of an animal, if bone strength is considered to be constant

It is this reasoning which prompted Galileo's attack on proportional design at the beginning of his book. Galileo extends his conclusions about the strength of beams of different sizes to any structure, either natural or artificial. The argument is today called the "square-cube law" and it is still considered by many engineers as an irrefutable demonstration of the impossibility of proportional design. (For a discussion of the influence of scale in structural design see [Aroca 1999].

The shadow of Galileo. We said in the previous section that architects and engineers have always used proportional rules in masonry architecture design. If we accept Galileo's argument all of these rules are incorrect. Galileo's conclusion has been accepted by many historians of engineering and has conditioned in a negative way the appreciation of the traditional proportional design rules. For example, Parsons said :

> There were no means of testing materials to determine their resistance to strain and consequently, the designer could not estimate the strength of a member nor did he have a theory by which he would compute the amount of strain that a member would be called to bear. There was, therefore, a *vicious circle of ignorance* [Parsons 1939] (my italics).

And Benvenuto wrote:

> ... il dimensionamiento in chiave geometrica restò sino a tempi recenti, il criterio più seguito dagli architetti: *il persistente pregiuduzio* che solo Galileo cominciò a smuovere, secondo il quale strutture geometricamente simili dovrebbero avere identiche proprietà statiche ... aveva condotto numerosi tratattisti a definire in linguaggio geometrico la figura delle volte [Benvenuto 1981] (my italics).

Robert Mark, commenting on the design of Hagia Sophia, remarks on the same argument: "Geometry did play a major role in their conceptual design [of Hagia Sophia]; however, as no less an observer than Galileo also commented, geometry alone can never ensure structural success" [Mark 1990]. Some authors wonder themselves how it was possible, with such an erroneous approach, that the great buildings of the past were built; Harold Dorn wrote "... It is a tribute to their skill that with this assortment of anthropomorphic analogies, qualitative generalizations, traditional arithmetical proportions, rules-of-thumb and an intuitive (and incorrect) arch 'theory', Renaissance builders erected magisterial and lasting structures" [Dorn 1970].

A contradiction. Dorn hinted at the heart of the problem. If we consider as correct Galileo's argument we face a contradiction:

The great master builders of the past used proportional design rules, which are essentially incorrect.

Using these rules they built the masterpieces of architecture and engineering of the past.

It is not reasonable to believe that the masterpieces of historical architecture, which have survived for centuries or millennia, were designed following an incorrect approach. So, perhaps, the matter should be reconsidered.

In what follows we will make a short outline of the fundamental aspects of masonry structural design.

The design of masonry arches and vaulted structures

The matter may be best discussed with reference to the fundamental element of masonry architecture: the arch. We have seen that in a voussoir arch in equilibrium (see fig. 1 above), the stones transmit a thrust and that this thrust must be contained within the arch, to obtain a set of compressive stress equivalent to the thrust. The line obtained by joining the points of application of the thrust in every joint (the locus of these points) is the *line of thrust*. To understand the concept it is only necessary to have some familiarity with the parallelogram of forces. Traditionally two approaches have demonstrated its usefulness in arch analysis: the first is to consider the equilibrium of a semi-arch; the second is the analogy with the statics of hanging chains and cables.

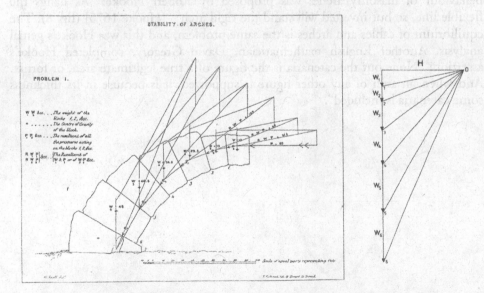

Fig. 7. *Left*, line of thrust in a semi-arch. The external horizontal thrust at the keystone is composed with the weights of the successive voussoirs defining a "path" of transmission of the forces [Snell 1846]. *Right*, the corresponding polygon of forces (added by the author)

Two half-arches: "A strength formed by two weaknesses". Consider the half-arch in fig. 7. It will clearly collapse, unless a force is applied in some point of the section at the keystone. If we apply an adequate horizontal thrust inside this section, then, as the figure shows, the thrust will be composed of the weight of every voussoir and the trajectory of the thrust will form the line of thrust drawn. In the original drawing the resolution of forces is made in the same drawing; however, as the horizontal thrust remains constant, all the lines may be gathered in one "force polygon" (added by the present author on the right of the figure). Then, using the terminology of graphical statics, the thrust line is an inverted funicular polygon, drawn tracing parallel lines to the force polygon. Note that in every joint the thrust is contained within the limits of the masonry. It is evident that by changing the value of the horizontal thrust we will obtain infinite lines of thrust within the arch. Also, we may change the point within the joint.

Now, we may imagine that we put another symmetrical semi-arch (a mirror reflection), on the other side. The horizontal thrust of the two semi-arches will equilibrate, no matter which line of thrust is considered: to use Leonardo's expression "an arch is a strength formed by two weaknesses". A semi-arch alone will collapse, but two "collapsing" semi-arches form a stable arch. It should be noted that the complete arch can be in equilibrium in infinite states of internal compression: in technical terms, the arch is statically redundant or "hyperstatic".

The cable analogy: "As hangs the flexible line . . .". Another way to understand the behaviour of masonry arches was proposed by Robert Hooke: "As hangs the flexible line, so but inverted will stand the rigid arch" [Hooke 1675] (fig. 8). The equilibrium of cables and arches is the same problem, and this was Hooke's genial analysis. Another English mathematician, David Gregory, completed Hooke's assertion: "None but the catenaria is the figure of a true legitimate arch, or fornix. And when an arch of any other figure is supported, it is because in its thickness some catenaria is included".

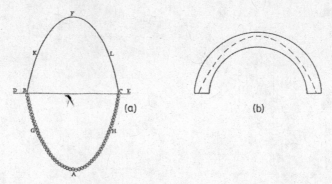

Fig. 8. (a) The arch as an inverted chain or cable. (b) The voussoirs of the arch may be imagined to be hanging from an imaginary chain, which represents the state of equilibrium. As the masonry must work in compression, the inverted "chain" must be contained within the arch [Heyman 1995]

The "material" masonry. The two previous analyses assume a material of certain properties, a "unilateral" material which can resist compression, but not tension. This condition "forced" the location of the thrusts within the masonry, to avoid tension. Traditional, unreinforced, masonry is just a pile of stones or bricks, disposed following some geometrical arrangement (the "bonding"), normally with mortar filling the joint or voids between the stones (sometimes there is no mortar). Roman concrete, where the volume of mortar is comparable to that occupied by the small stones, is also a type of masonry, and so is pisé (stiff earth or clay rammed until it becomes firm). In fact, any combination of stone, bricks, mortar or earth can lead to a successful kind of masonry. In fig. 9a, taken from a building handbook of ca. 1900, some types of masonry have been drawn. But in a common building we can find maybe a dozen different kinds of masonry (fig. 9b). Many times a structural element is a combination of several different masonries; this is typical of thick masonry walls with two external ashlar "shells" and a core made of rubble. Again, the number of combinations is almost unlimited.

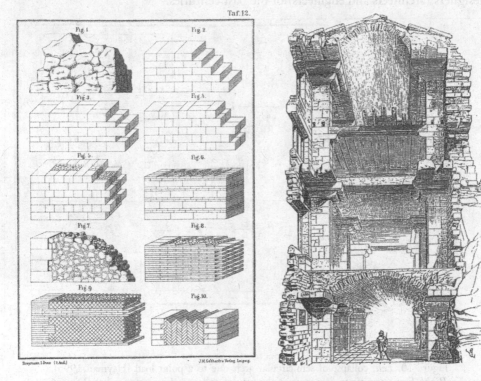

Fig. 9. (a) Different kinds of masonry [Warth 1903]. (b) Internal structure of a medieval building [Viollet-le-Duc 1854]

The question is: Where is the isotropic, homogeneous, elastic material of the classical elastic theory of structures? Nowhere. Masonry is essentially anisotropic, discontinuous, and heterogeneous; as for the elastic constants, one may ask, where?

In the external ashlar stone or in the rubble filling or in the mortar? It is easy to test a stone specimen to obtain, for example, the crushing strength. At the end of the nineteenth century thousands of tests were made on stone, brick and mortar. But when it came to ascertain the strength of the material "masonry" only vary vague indications, with enormous safety coefficients, were given (for stone tests see for example [Debo 1901]; for a critical discussion of the matter see [Huerta 2004]).

The essential characteristic of masonry is that it has good compressive strength and almost no tensile strength. Also, the stones maintain their position (no sliding occurs) due to the high friction coefficient (ca. 0.5). Heyman [1966, 1995] has systematized these observations into three Principles of Limit Analysis of Masonry: 1) Masonry has an infinite compressive strength; 2) Masonry has no tensile strength; 3) Sliding is impossible. These principles are reasonable and easy to check; they have been accepted, implicitly or explicitly, by all the masonry designers (architects and engineers) of the past centuries.

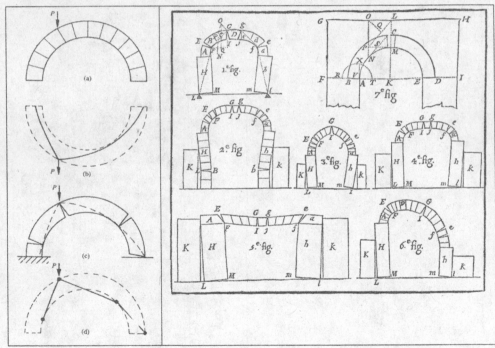

Figure 10. *Left*, collapse of semicircular arch due to a point load [Heyman 1995]. *Right*, first attempts to demonstrate the way of collapse of masonry arches [Danyzy 1732].

Collapse of masonry arches. If the material has these properties (and most types of masonry do), then the analysis of masonry structures may be included within the frame of Limit Analysis (or plastic theory) as Professor Heyman showed in 1966. The proof is complex, but the essence of the matter lies in that for a material of these properties, collapse will only occur when a sufficient number of hinges form which converts the structure into a cinematically admissible mechanism. The "hinges" are the points where the line of thrust touches the limit of the masonry.

Fig. 10, left, shows the process of collapse of an arch which supports an increasingly growing point load: the form of the "hanging chain" is modified until there is only one line of thrust inside the arch, touching alternatively the intrados or intrados; at the bottom the corresponding four-bar collapse mechanism is shown. The same applies to more complex structures. As a matter of interest, on the right side of the same figure appear the first essays made by the French scientist and engineer Danyzy in 1732, which demonstrate this mode of collapse.

The Safe Theorem. Now, within the frame of Limit Analysis, it can be demonstrated that if it is possible to draw a line of thrust within an arch, then this arch is safe, i. e., it will not collapse. This is a corollary of the Safe Theorem (or Lower Bound Theorem) of Limit Analysis. The theory of Limit Analysis, also called the Plastic Theory, was developed during the 1930s–1950s and it constitutes the fundamental contribution to the theory of structures in the twentieth century (in the same way as Elastic Theory was the main contribution in the nineteenth century). Professor Heyman has shown that the Fundamental Theorems, originally derived for steel frames, can be translated to masonry structures. His work has put the theory of masonry structures within the frame of the modern theory of structures; he has exposed with great clarity and intellectual rigour the consequences of this translation. The theory of plasticity itself is difficult, but, as sometimes occurs, the consequences are quite easy to understand. For example, the condition that the line of thrust must be contained within the arch leads to purely geometrical statements.

The limit arch. An arch of sufficient thickness will contain infinite lines of thrust, for example the arch in fig. 11a. If we reduce the thickness of the arch, the form of the line of thrust will suffer no change, but it is evident that for a certain thickness only one line will be contained within the arch: this arch is the *limit arch* and its thickness is the *limit thickness* (fig. 11b).The limit thickness can be expressed conveniently as a certain fraction of the span. For a semicircular arch the limit thickness is nearly 1/18 of the span. That means that a masonry arch thinner than this proportion cannot be built; the arch will become a mechanism, collapsing (fig. 11c). Thus, the limit arch forms the point of departure for the design of a safe arch: we will obtain geometrical safety by "thickening" the limit arch.

There are two approaches for the design of a safe arch: the "strength" approach and the "stability" approach. In both approaches the limit arch is the point of departure. If we want to follow a condition of strength, the thickness should be increased until the stresses reach a certain "admissible" value obtained dividing the crushing stress by a certain coefficient. If we are concerned with a possible failure by lack of stability (the formation of a collapse mechanism), then we increase the thickness multiplying it by a certain *geometrical factor of safety* (this concept has been introduced by [Heyman 1969]). In the case of arches, a typical value is 2 or 3; so a safe arch will have double the thickness of the limit arch.

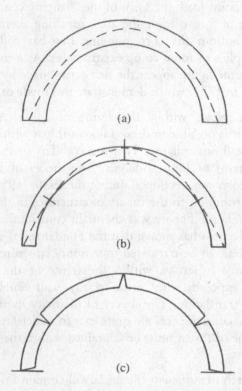

Fig. 11. (a) An arch of sufficient thickness contains comfortably infinite lines of thrust. One has been drawn. (b) Diminishing the thickness, we arrive at the limit thickness, where only one line can be drawn within the arch. (c) At t he points where the line touches the limit of the masonry a "hinge" forms. The arch is in mathematical equilibrium [Heyman 1995]

Masonry arch design: Strength versus Stability. The problem is which condition governs the design. Maybe the best thing is to take an example. Consider an arch of stone of 18m span. The limit arch will have a thickness of very nearly 1m.

Strength: if the arch is going to be made of medium sandstone (20 kN/m^3) with an admissible working stress of, say, 4 N/mm² (1/5 of a crushing strength of 20 N/mm²). Then, it is easy to calculate (considering a uniform stress distribution) that the required increase of thickness will be of 80mm or 0.5% of the span. (In fig. 12b the increase of thickness has been exaggerated: in fact the increase will be within the thickness of the lines of intrados and extrados in fig. 12a.) The arch is so near the limit thickness that it is on the verge of collapse; in fact, by inspection, the overall form of the arch has not changed.

Stability: the usual geometrical factor of 2 or 3 (the last value represented in fig. 12c) will impose a substantial change in the form of the arch, which will be easily recognised by inspection. Any master mason will know, just seeing it, that the arch is not only safe, but that it has a surplus (the usual geometrical factor of safety for arches is 2).

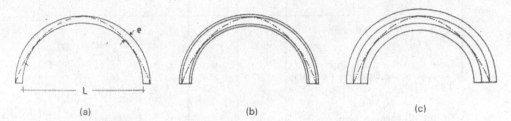

Fig. 12. Design of a masonry arch. (a) The limit arch; (b) Design by "strength"; the increase of thickness should guarantee admissible stresses; (c) Design by "stability"; the increase of thickness should afford a geometrical safety to the arch

The strength criterion is unsafe: the stresses will be low but the arch will be dangerously near of the collapse situation. Of course, the calculations have been made for an arch of 18m. For greater arches the stresses will grow linearly, and the increase of thickness will be correspondingly greater.

Limit spans for masonry. For a certain span, the thickness by strength will coincide with the thickness by stability and this point will mark the limit of the span of the arch. For a geometrical factor of safety of 2 (and considering a rectangular block of stresses at the base) this maximum span will be:

$$s_{max} = \frac{2}{\pi} \frac{\sigma_{adm}}{\gamma}$$

where σ_{adm} is the maximum admissible working stress for the material and γ is the specific weight of the masonry. For the above stated data, s_{max} = 128m and the absolute maximum span for the crushing strength will be 5 times this span, or 640m. With stones of a better quality, correspondingly greater spans may be built. The dimensions are well above the usual dimensions of bridges. The largest span of an stone arch bridge is that of Fong-Huan, in China, built in 1972, with 120m [Fernández Troyano 1999]. In concrete (with no longitudinal reinforcement) it is the bridge of Caille in Cruseilles (1928) with 139.8m. The quantity σ_{adm}/γ, which is a length, represents the limit height of a column of uniform section built with this material; this quantity was used in the nineteenth century as a measure of the strength of the materials and, also, as an indication of the maximum sizes which can be attained (fig. 13).

INDICATION DES MATÉRIAUX.	Poids du décimètre cube.	Charge d'écrasement par centimètre carré.	Hauteur représentative de la charge d'écrasement (1).	OBSERVATIONS.
Pierres volcaniques.	kilogr.	kilogr.	mètres.	
Basalte de Suède.	3,06	1912	6248	Rondelet.
Basalte d'Auvergne.	2,88	2078	7215	Id.
Lave du Vésuve, dite *Piperno.* . .	2,60	563	2165	Id.
Lave grise des environs de Rome.	1,97	228	1157	Id.
Tuf de Rome.	1,22	58	478	Id.
Granits.				
Granit d'Aberdeen bleu.	2,62	767	2927	G. Rennie.
Granit vert des Vosges..	2,85	620	2175	Rondelet.
Granit gris de Bretagne.	2,74	654	2383	Id.
Granit de Normandie, Gatmos. . .	2,66	702	2639	Id.
Granit gris des Vosges.	2,64	423	1603	Id.
Grès.				
Grès très-dur.	2,52	813	3226	Id.
Grès blanc.	2,48	923	3713	Id.
Grès bigarré des Vosges.	2,17	400	1843	Conservatoire des arts et métiers.
Pierres calcaires.				
Marbre noir de Flandre..	2,72	789	2901	Rondelet.
Marbre blanc veiné.	2,70	298	1104	Id.
Marbre rouge du Devonshire.. . .	2,70	522	1933	Rennie.
Calcaire de Portland..	2,42	262	1083	Id.
Pierre de Caserte, près Naples. . .	2,72	595	2191	Rondelet.
Pierre noire de St-Fortunat (Lyon).	2,65	627	2366	Id.
Liais de Bagneux, près Paris. . .	2,44	445	1824	Id.
Travertino de Rome.	2,36	298	1262	Id.
Roche de Châtillon, près Paris. . .	2,29	174	760	Id.
Roche douce de Châtillon.	2,08	134	644	Id.
Roche d'Arcueil, près Paris. . . .	2,30	253	1100	Id.
Pierre de Saillancourt, 1ʳᵉ qualité.	2,41	141	585	Id.
Briques.				
Brique dure très-cuite.	1,55	150	962	
Brique rouge.	2,17	57	262	
Brique rouge pâle.	2,08	39	187	
Mortiers.				
Mortier de chaux et de sable de rivière.	1,63	31	»	Rondelet.
Mortier de ciment de tuileau. . .	1,46	48	»	Id.
Mortier de pouzzolanes de Naples et de Rome mêlées.	1,46	37	»	Id.
Mortier avec chaux éminemment hydraulique.	»	144	»	Vicat.

(1) Cette colonne indique la hauteur du prisme droit de la matière considérée dont le poids serait suffisant pour écraser sa propre base (§ 25).

Fig. 13. Table of the strength of stones and bricks. On the second column from the right, the limit height in meters has been calculated [Collignon 1885]

Engineers of the past were well aware of the possibility of building great spans in masonry. The bridge over the Adda in Trezzo, built in 1370–77 (demolished in 1416 for military reasons) had a span of 72 m (this span was surpassed only after 1900). But perhaps the best example of the confidence that enormous spans can be made is found in a design of Leonardo da Vinci, ca. 1500, for a bridge of one arch over the Golden Horn in Istanbul (fig. 14, left). Leonardo's bridge would have had a span of 240 m and in the manuscripts he shows concern only for the problems of centering. Stüssi [1953] undertook an exhaustive analysis of Leonardo's design and concluded that it would have been feasible. Stüssi obtained a maximum stress of 10 N/mm² for a material with a specific weight of 28 kN/m³ (fig. 14, right).

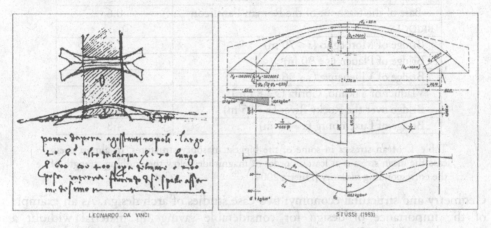

Fig. 14. *Left*, Leonardo de Vinci's design for an arch bridge with a span of 240 m, over the Golden Horn in Istanbul (ca. 1500). *Right*, Stüssi's analysis.

Stresses in masonry buildings. In buildings the same thing occurs. Even in the greatest structures built the calculated mean stresses in the most loaded parts (in general the columns) are still quite moderate, as can be seen in Table 1 below. For example, in the main piers of St. Peters, which supports a dome and drum with a total weight of 400,000 kN, the mean stress is 1.7 N/mm². A similar structure three times bigger could have been built, but would it have had any meaning?

In conclusion, it is a fact that for historical masonry structures, the stresses are an order or two orders of magnitude below the crushing strengths of the masonry and, therefore, the problem of masonry design is not governed by strength but by stability.

Stability governs the design, which means that the objective is to design safe formsl. Considerable savings of material may be obtained by choosing adequate "geometries". Economy was, of course, the second main structural concern of old master builders. Until recent times, the search for economical structures and economical building procedures has been a constant. Choisy showed his surprise when he discovered in 1873 that precisely this striving for economy was the key to a deeper understanding of the Imperial Roman building processes.

BUILDINGS	Mean stress N/mm²
Columns, church of Toussaint d'Angers	4.4
Main pillars, French Pantheon (St. Genevieve), Paris	2.9
Main pillars, Hagia Sophia	2.2
Main pillars, cathedral of Palma de Mallorca	2.2
Main pillars, St. Paul, London	1.9
Main pillars, St. Peters. Rome	1.7
Main pillars, Church des Invalides , Paris	1.4
Main pillars in the cathedral of Beauvais	1.3
Base of the tambour of the Roman Pantheon	0.6
BRIDGES	
Bridge of Morbegno ($s = 70$ m)	7.0
Bridge of Plauen ($s = 90$ m)	6.9
Bridge of Villeneuve ($s = 96$ m)	5.7
Viaduct of Salcano, Göritz ($s = 85$ m)	5.1
Bridge over the Rocky River ($s = 85$ m)	4.4
Bridge of Luxemburg ($s = 85$ m)	4.8

Table 1. Mean stresses in some of the biggest masonry structures. In almost every case the mean stress is at least an order of magnitude below the crushing strength of the corresponding masonry [Huerta 2004]

Geometry and structural economy: two case studies of arch design. As an example of the importance of design for considerable savings in material without a diminution of the safety, a simple case will be investigated: the design of a simple barrel vault on rectangular buttresses. The first thing is to design the vault with an adequate geometrical safety. For a semicircular arch the limit thickness is nearly $t = 1/18$. But for segmental arches with an opening angle of less than 180° the limit thickness diminishes very rapidly (fig. 15, left).

The design strategy to use "thin" vaults safely is to reduce the height of the arch, filling the haunches of the vault with good masonry up to a certain height. In fig. 15, above, two designs have been drawn of vaults with the same geometrical safety factor of 3 (thickness 3 times the limit): a good filling of the haunches up to half the height of the vault permits a reduction of the division of the thickness of the vault by 4. The thrust of the vault is, therefore, also divided by nearly the same factor and the buttresses (for a height equal to the span) may be reduced from nearly 1/3 of the span to nearly 1/4 of the span. The total amount of masonry (vault plus buttresses) is reduced by 40%. This remarkable economy is the result of just putting some rubble masonry between the haunches and the wall.

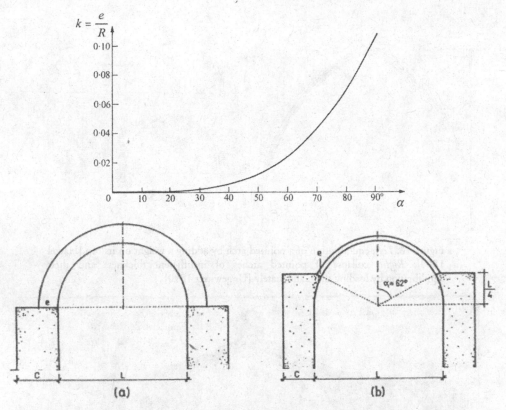

Figure 15. *Above,* limit thickness of circular segmental angles, depending on the opening semi-angle [Heyman 1995]. *Below,* effect of the reduction of the opening angle by filling the haunches with good masonry [Huerta 2004].

Another example of the same kind of geometrical design is that of pointed arches. A pointed arch has, in general, a smaller limit thickness than the corresponding semicircular arch of the same span. However, a simple filling of the haunches will not reduce the limit thickness with the same rapidity as in a semicircular arch, due to the solution of continuity at the top. In fact, to achieve good results a pointed load is needed at the tip of the arch. This load will "break" the smoothness of the line of thrust, better adapting it to the form of the arch. In the drawing by Hatzel (fig. 16, left) on the left side the line of thrust has been drawn; it is evident that it will be impossible to introduce it within the arch whose thickness is less than the limit thickness. However, on the right side a weight has been added on top; now the line of thrust coincides almost perfectly with the middle line and goes out in the lower half, where some filling should be put on the haunches to allow the thrust pass to the buttress system.

load on the keystone [Huerta 2004]. Ardemgan was also well-aware of this when explaining the stability of medieval arches, in his 1890 edition of Ungewitter's manual and his drawings explain it very well (see section 7) (fig. 16, right).

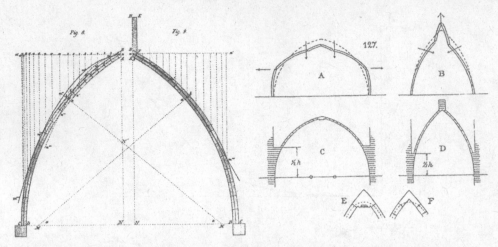

Figure 16. *Left*, equilibration of a pointed arch by adding a weight on its top [Hatzel 1849]. *Right*, collapse of pointed arches of insufficient thickness and their equilibration by loading them adequately [Ungewitter 1890]

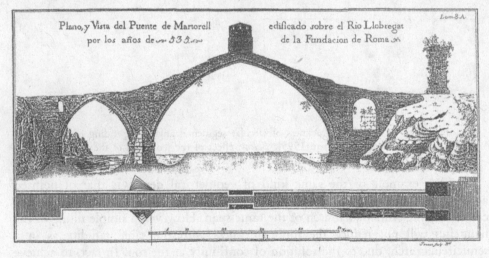

Fig. 17. Puente del Diablo (devil's bridge) in Martorell; thirteenth century, on Roman springings. The tower at the top of the bridge helps to stabilise the thin ring of the arch (after [Sánchez Taramas in Muller 1769])

This was well known by the Gothic master builders and the heavy, sometimes richly sculpted, keystones of pointed arches and cross vaults serve a structural as well as a decorative function. Tosca, commenting on the design of pointed medieval arches, said: "These arches correspond to the Gothic order, and though being beautiful, they are weak near the haunches … particularly if they have no load on the keystone" [Tosca 1707]. Mohrmann was also well aware of this when explaining the statics of medieval arches in his 1890 edition of Ungewitter's manual and his drawings explain clearly Tosca's assertion (fig. 16, right).

Bridge builders were also aware of this. The medieval Puente del Diablo (devil's bridge) in Martorell shows a little tower on the keystone (fig. 17). The construction serves to control the passage through the bridge but it also plays a fundamental role in stabilising the thin ring of the arch. If, for example, during some work of restoration the tower were to be removed to be re-built afterwards, this may prove to be a dangerous operation.

The "ideal" arch. Now we may make a short digression about the "ideal" form of the arch. Is it the catenary, as was maintained by Hooke and Gregory, and later by many others up until the present day? In a catenarian arch all the voussoirs are different as the radius of curvature changes from point to point. It is difficult to build the centering and if the arch is made of stones, every stone will need a different template. The Gothic approach is cleverer: you choose a geometrical form made of circular arcs and then you load this arch to make it "catenarian". The rubble filling and the stone at the top have almost no cost. A "catenarian" architecture like that of Gaudí is, in fact, quite expensive in comparison with Gothic architecture. We will leave the matter here, because to pursue it we must enter in the realm of architectural design which, of course, involves many other aspects besides that of structural efficiency (for a discussion of Gaudí's structural design see [Huerta 2003]).

A wonder of equilibrium: the cathedral of Palma de Mallorca. Up to now we have discussed just simple examples of the kind of geometrical design which is characteristic of masonry architecture. Of course, in a building of some scale many different structural problems are present and the master builder must take into account all of them and, eventually, produce an integrated design. The degree of subtlety which can be found in some projects is amazing. A good example is that of the cathedral of Palma de Mallorca (fig. 18).

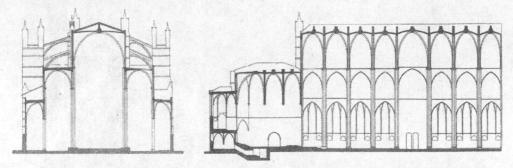

Fig. 18. Cross and longitudinal section of the cathedral of Palma de Mallorca [Domenge 1999]

This is one of the biggest Gothic cathedrals: the main nave has a span of 20m and a height of 42m. The nave columns are extraordinary slender and must support at the top the thrust of the lateral aisles. How is it possible for such a slender column to function as a buttress?

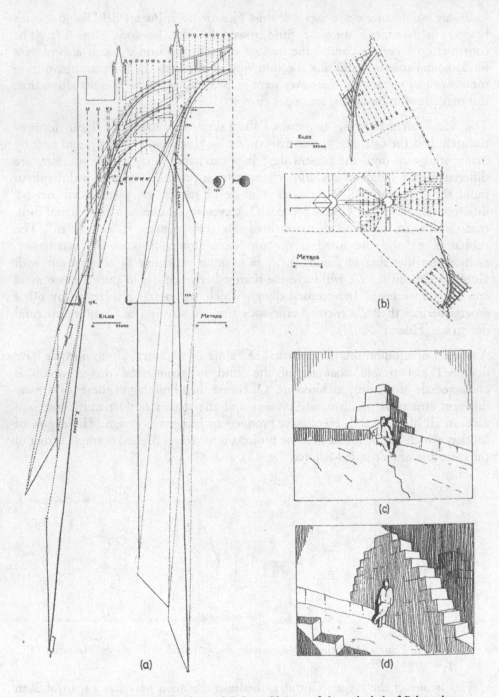

Fig. 19. (a) Static analysis of the global equilibrium of the cathedral of Palma de Mallorca's cathedral. (b) Statical analysis of the vault's thrusts [Rubió i Bellver 1912]; (c) and (d) Loads on top of the crossing and transverse arches (redrawn by the author after [Rubió i Bellver 1912])

The question cannot be answered without climbing to the extrados of the high vaults. There one may see pyramids of stone on top of the crossing arches, heavy transverse walls on the transverse arches, extraordinarily thick ribs, etc. The unknown master used a typical Gothic device: buttressing by loading. The weight on top of the columns is increased extraordinarily so that the thrust of the lateral aisle only deviates slightly the vertical direction of the loads. Of course, these extraordinary loads produce a lateral thrust and the external buttresses which receive it are the biggest of all Gothic architecture, with a depth (8m) approaching one half of the span. In 1912 Rubió i Bellver, a disciple of Gaudí, published a static analysis of the cathedral and this analysis confirms the qualitative comments made before. In Rubio's drawing of the trajectory of the thrusts, with the different forces drawn to scale, it is easy to see the equilibrating effect of the added loads (fig. 19).

We have considered the problem of masonry arch design in some detail in order to show the importance of the geometry on the arch's safety. The example of Palma de Mallorca illustrates how this kind of geometrical design is at the heart of masonry design. This depends on the Safe Theorem of Limit Analysis which, as Professor Heyman has said, is "the rock on which the whole theory of structural design is now seen to be based" [Heyman 1999]. The main corollary of this theorem leads to the "approach of equilibrium" and, for a masonry structure, the problem of obtaining a compressive state of equilibrium is a geometrical problem. Citing again Heyman: "The key to the understanding of masonry is to be found in a correct understanding of geometry" [1995].

Conclusion: The "error" of Galileo and "Navier's straitjacket"

Galileo was the first to provide a theory which permitted the strength of a certain type of structural element – the simple beam – to be checked. He was the founder of the theory of structures. He was right in deducing that the strength of a certain section of a beam is proportional to the strength of the material and to the area and depth of its cross-section. Also correct is his observation that, given any structure which supports its own weight, if we multiply its "size" by a certain factor, maintaining the geometrical form of the structure, the loads grow with the cube of the factor, but the sections of the structural members grow with the square, and as their strength are proportional to the areas, either the structure becomes "weaker" or the members must be thickened. In modern terms: stresses grow linearly with the dimensions, that is, they are directly proportional to the scale factor. All this is completely correct and it is an extraordinary feat of genius that Galileo, working alone in his old age, not only founded the New Science of the Strength of Materials, but that also drew design conclusions: the square-cube law.

Galileo realized that his discovery contradicted a traditional design approach, or rule: that of proportional design. He was well aware that it was an important discovery, which would affect the design of many types of structures: machines, ships, beam and framed structures. The theory also permitted an explanation of some biological facts: why the bones of small animals are proportionally slender and, also, why small animals are proportionally stronger. He expressed his discovery so convincingly that his argument has reached the rank of "Law" in the books on structural design: "the square-cube law".

The law refers only to problems of the design of structures supporting mainly their own weight when the governing criterion is strength. This is, indeed, one of the three fundamental structural criteria: strength, deformation and stability. A structure should resist its loads without the breaking of any of its members. It also should not present unduly large deformations. Finally, the structural elements and the structure as a whole should not be unstable.

In modern structures strength is usually the governing criterion. But, as we have seen, in historical constructions strength plays no role and it is stability which is relevant. Galileo was perhaps too quick to generalize his discovery to any structure, including specifically "palaces or temples". He was wrong in applying his "strength" argument to masonry structures. But it is understandable that a scientist should look for universal laws. It is also understandable that he should be carried away by enthusiasm for a great discovery.

But after more than three centuries it is remarkable that Galileo's argument continues to be applied uncritically to structures where it is evident, in the etymological sense, that it does not apply, by simple comparison of structures of different sizes. In fact, any reader of books on the history of architecture would have problems in ascertaining the actual size of a building from a plan without scale. If we compare the form of the domes of San Biaggio in Montepulciano, St. Peters in Rome and Santa Maria del Fiore, drawing the three at the same size, we may see that the overall form and proportions are very similar (fig. 20).

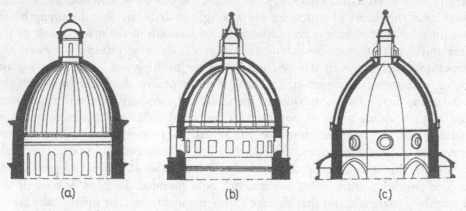

Fig. 20. Comparison of the form of three Renaissance domes: (a) San Biaggio in Montepulciano (14m); (b) St. Peters in Rome (42m); (c) Santa Maria del Fiore (42m). Note that, although the first dome is three times as small than the other two the form is very similar

However, the dome of San Biaggio has a span of 14m, 1/3 of the 42m of both St. Peters and Santa Maria del Fiore. Byzantine architecture is full of domes similar to that of Hagia Sophia and the geometry of cupola of the Pantheon has been reproduced hundreds of times.

Now the question is why an argument which doesn't coincide with the facts (the design of masonry architecture) has been used for three centuries and continues to be used. The first reason has been already mentioned: the argument *is true*. For any structure subject to its own weight, including masonry structures, internal stresses grow linearly with size in similar structures. What is not true is that stresses (strength) determine the design of masonry structures.

The emphasis on strength and the opposition of the "actual" internal stresses in structures as the main objective of structural theory comes from the very development of the science of structures. Navier in his *Leçons* of 1826 stated explicitly the aims of the structural theory: the resolution of the three structural equations, those of equilibrium, of the elastic material and of compatibility, will give a unique solution for the internal forces within the structure. Then internal stresses will be calculated from them and, finally, these stresses will be compared with values of the strength of the material obtained in experimental tests. The focus is on obtaining the stresses and the complicated mathematical apparatus of the theory of elasticity and of the resolution of the system of equations precluded for almost one hundred years any criticism. Professor Heyman has called this frame of reference "Navier's straitjacket" [1999]. Nowadays the Method of Finite Elements, the numerical resolution of the system of the three structural equations dividing the structure in "finite elements", points in the same direction as the "old" classical elastic theory. Sometimes, of course, this has negative consequences in the field of structural intervention on historical buildings. As we have seen, many times the only way to assure the safety of a building is to "overload" some of its parts, as it occurred with the main columns in the cathedral of Palma de Mallorca. A reduction of weight, which "theoretically" always leads to a reduction of stresses, may lead to serious damage and, eventually, to the collapse of the structure.

In summary, any engineer or architect with some formation in structural theory feels more comfortable within the frame of the strength approach of Galileo and the classical theory of structures. It requires an effort, and some study, to overcome our own prejudices and to accept that, for example, the medieval master masons, knowing nothing of mathematics, elastic theory and strength of materials, had a deeper understanding of masonry architecture than we engineers and architects of the twenty-first century do. However the masonry buildings of the past stand today as a proof of their knowledge and our ignorance.

References

AROCA HERNÁNDEZ-ROS, Ricardo. 1999. *¿Qué es estructura?* Madrid: Cuadernos del Instituto Juan de Herrera.

BENVENUTO, Edoardo. 1981. *La Scienza delle Construzioni e il suo sviluppo storico.* Florence: Sansoni.

BESENVAL, Roland. 1984. *Technologie de la voûte dans l'Orient Ancien.* 2 vols.Paris: Editions Recherche sur les Civilisations.

COENEN, Ulrich. 1990. *Die spätgotischen Werkmeisterbücher in Deutschland. Untersuchung und Edition der Lehrschriften für Entwurf und Ausführung von Sakralbauten. (Beiträge zur Kunstwissenschaft, Bd. 25).* München: Scaneg.

COLLIGNON, Édouard. 1885. *Cours de mécanique appliquée aux constructions. Première partie: Résistance des matériaux.* 3rd. ed. Paris: Vve. Ch. Dunod.

DEBO, Ludwig. 1991. *Lehrbuch der Mauerwerks-Konstruktionen.* Hannover: Gebruder Jänecke.

DOMENGE I MESQUIDA, Joan. 1999. La catedral de Mallorca: Reflexiones sobre la concepción y cronología de sus naves. In: *Gotische Architektur in Spanien,* ed. C. Freigang, 159–187, 395–398. Frankfurt a. M./Madrid: Vervuert/Iberoamericana.

DORN, Harold I. The Art of Building and the Science of Mechanics. A Study of the Union of Theory and Practice in the Early History of Structural Analysis in England. Ph.D. thesis, Princeton University, 1970.

DURM, Josef. 1885. *Die Baukunst der Etrusker und Römer. (Handbuch der Architektur. Zweiter Teil. 2.Band).* Darmstadt: J. Ph. Diehl.

EL-NAGGAR, Salah. 1999. *Les voûtes dans l'architecture de l'Égypte ancienne.* 2 vols. Cairo: Institut Français d'Archéologie Orientale.

FERNÁNDEZ TROYANO, Leonardo. 1999. *Tierra sobre el agua. Visión histórica universal de los puentes.* Madrid: Colegio de Ingenieros de Caminos Canales y Puertos.

GALILEI, Galileo. 1638. *Discorsi e Dimostrazioni Matematiche intorno à due nuove sicenze Attenenti alla Mecanica & i movimenti Locali.* Leiden: Elsevier. (English translation by Henry Crew and Alfonso de Salvio, New York: Dover, 1954.)

GREGORY, D. 1697. Catenaria. *Philosophical Transactions of the Royal Society* 19, 231: 637–652.

HATZEL, E. 1849. Über die Technik in spezieller Beziehung auf die Architektur und die Gestaltung der Formen. *Allgemeine Bauzeitung,* 132–169.

HECHT, Konrad. 1979. *Maß und Zahl in der gotischen Baukunst.* Hildesheim: Georg Olms Verlag.

HEYMAN, Jacques. 1966. The Stone Skeleton. *International Journal of Solids and Structures* 2: 249–79.

———. 1998. *Structural analysis: a historical approach.* Cambridge: Cambridge University Press.

———. 1995. *The Stone Skeleton. Structural Engineering of Masonry Architecture.* Cambridge: Cambridge University Press.

———. 1999. *The Science of Structural Engineering.* London: Imperial College Press.

———. 1999. Navier's straitjacket. *Architectural Science Review* 42: pp. 91-95.

HOOKE, Robert. 1675. *A description of helioscopes, and some other instruments.* London.

HUERTA, Santiago. 2003. El cálculo de estructuras en la obra de Gaudí. *Ingeniería Civil* 130: 121–33.

———. 2004. *Arcos, bóvedas y cúpulas. Geometría y equilibrio en el cálculo tradicional de estructuras de fábrica.* Madrid: Instituto Juan de Herrera.

———. 2006. Geometry and equilibrium: The gothic theory of structural design. *Structural Engineer* 84: 23–28.

KOEPF, Hans. 1969. *Die gotischen Planrisse der Wiener Sammlungen.* Vienna: Hermann Böhlaus Nachfolger.

MARK, Robert. 1990. *Light, wind, and structure.* Cambridge, MA: MIT Press.

MULLER, Juan. 1769. *Tratado de Fortificación ó Arte de construir los Edificios Militares, y Civiles ... traducido y aumentado con notas de D. Miguel Sánchez Taramas.* Barcelona: Thomas Piferrer.

NAVIER, C.L.M.H. 1826. *Résumé des Leçons donées à L'École des Ponts et Chaussées, sur l'application de la mécanique à l'établissement des constructions et des machines.* Paris. (2nd. ed. 1833; 3rd ed. with notes and appendices by B. de Saint-Venant 1864.)

PARSONS, W.B. 1976. *Engineers and Engineering in the Renaissance.* Cambridge, MA: MIT Press. (1st ed. 1939.)

RUBIÓ I BELLVER, Joan. 1912. Conferencia acerca de los conceptos orgánicos, mecánicos y constructivos de la Catedral de Mallorca. *Anuario de la Asociación de Arquitectos de Cataluña* (1912): 87–140.

SNELL, George. 1846. On the Stability of Arches, with practical methods for determining, according to the pressures to which they will be subjected, the best form of section, or variable depth of voussoir, for any given extrados or intrados. *Minutes and Proceedings of the Institution of Civil Engineers* 5: 439–476, Plates 27–40.

TOSCA, Tomás Vicente. 1707–15. *Compendio mathemático en que se contienen todas las materias más principales de las ciencias que tratan de la cantidad …* Valencia: Antonio Bordazar.

UNGEWITTER, Georg Gottlieb. 1890. *Lehrbuch der gotischen Konstruktionen. III Auflage neu bearbaitet von K. Mohrmann.* 2 vols. Leipzig: T.O. Weigel Nachfolger.

VIOLLET-LE-DUC, Eugene. 1854–1868. *Dictionnaire raisonnée de l'Architecture Française du XI au XVI siécle.* 10 vols. Paris: A.Morel.

WARTH, Otto. 1903. *Die Konstruktionen in Stein.* Leipzig: J. M. Gebhardt's Verlag.

About the author

Santiago Huerta became an architect in 1981 following study at the School of Architecture of the Polytechnic University of Madrid. He was in professional practice from 1982 to 1989. In 1989 he became Assistant Professor in the School of Architecture of Madrid. He earned a Ph.D. in 1990 with a dissertation entitled "Structural design of arches and vaults in Spain; 1500-1800". Since 1992 he has been Professor of Structural Design at the School of Architecture of Madrid. In 2003 he became President of the Spanish Society of Construction History. From 1992 until the present he has been a consulting engineer for the restoration of many historical constructions, including the Cathedral of Tudela, San Juan de los Reyes in Toledo and the Basílica de los Desampardos among others, as well as some medieval masonry bridges. Since 1983 his research has focused on arches, vaults and domes, masonry vaulted achitecture in general. He is the author of *Arcos, bóvedas y cúlulas. Geometria y equilibrio en el cálculo tradicional de estructuras de fábrica* (Madrid: Instituto Juan de Herrara, 2004).

Paul Calter

108 Bluebird Lane
Randolph Center, VT
05061 USA
pcalter@sover.net

keywords: arch, St.
Louis Arch, Eero
Saarinen, Gateway to
the West, Gateway
Arch, catenary,
parabola, hanging chain

Research

Gateway to Mathematics
Equations of the St. Louis Arch

Abstract. Eero Saarinen's Gateway Arch in St. Louis has the form of a catenary, that is, the form taken by a suspended chain. The catenary can be reproduced empirically, but it can also be precisely formulated mathematically. The catenary is similar to the paraboloid in shape, but differs mathematically. Catalan architect Antoni Gaudi used the catenary to great effect in his Church of the Sagrada Familia in Barcelona, but he also used the paraboloid as well.

An arch consists of two weaknesses which, leaning one against the other, make a strength.

Leonardo da Vinci

Introduction

When building a circular masonry arch, a builder was likely to construct a wooden form on which the stones or bricks were laid, the shape being traced with pegs and string or a radius bar. But for larger arches, and those made of steel, fabricated at a steelyard and assembled on-site, it is useful to work from an *equation*. If fact, equations of geometric figures like the circle were not even available until René Descartes developed analytic geometry in the seventeenth century.

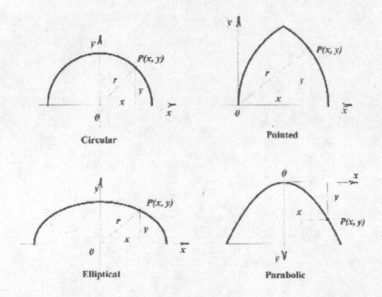

Fig. 1.

An equation of the arch will give the height y of any point on the arch at a given horizontal distance x. Both distances are measured from chosen axes. An equation is, of course, meaningless unless one knows the x and y axes, and the origin, the point where they intersect.

For example, here are the equations for the circular, pointed, parabolic and elliptical arches shown in fig. 1, if the axes are placed as shown. Changing the position of an axis would change the equation.

Circular:
$$y = \sqrt{r^2 - x^2}$$

One side of Pointed:
$$y = \sqrt{r^2 - x^2}$$

Elliptical:
$$y = b\sqrt{1 - \frac{x^2}{a^2}}$$

Parabolic:
$$y = kx^2$$

Note that for the parabola, we have taken the origin at the vertex of the curve, and the positive y axis vertically downward.

Turning to the St. Louis arch, note that we have three curves, for the intrados, the extrados, and the centerline (fig. 2). What I call the centerline is the curve connecting the centroids of the cross-section.

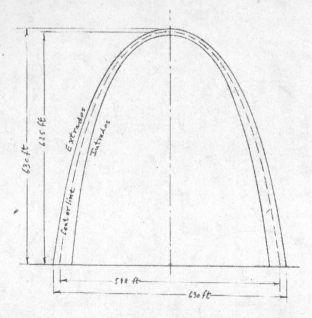

Fig. 2.

The cross-section of the arch is an equilateral triangle, whose centroid is at the intersections of the three medians (fig. 3). For an equilateral triangle, that same point is also the orthocenter (the intersections of the altitudes) the incenter (intersection of the angle bisectors) and, for good measure, the circumcenter (intersection of the perpendicular bisectors of the sides.)

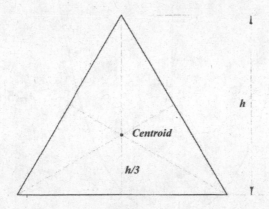

Fig. 3.

The Catenary

The published equation of the centerline of the St Louis arch, with the constants rounded to three digits, is

$$y = 68.8(\cosh 0.01 - 1)$$

Let us break it down. First, what is a *cosh*? It is one of the six hyperbolic functions. The hyperbolic functions are analogous to the circular functions, but are based on the hyperbola rather than the circle. In particular, the hyperbolic cosine is defined as

$$\cosh ax = \frac{e^{ax} + e^{-ax}}{2}$$

So with $a = 0.01$, our arch equation becomes

$$y = 68.8\left(\frac{e^{0.01x} + e^{-0.01x}}{2} - 1 \right)$$

Fig. 4 shows the graph of that equation. Notice that the curve opens upwards, but we can correct that simply by inverting the curve. We then take the origin at the high point, with the y axis downward and along the centerline of the curve.

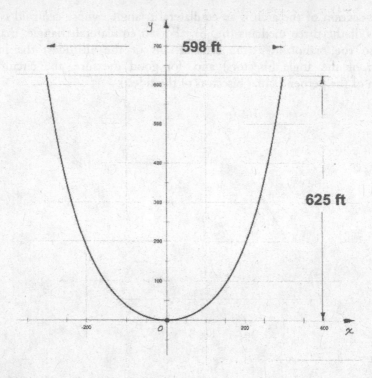

598 ft

625 ft

Fig. 4.

The constants in the arch equation, 68.8, 0.010, 2, and −1, serve to scale the curve to the proper height and width, and to shift the curve vertically. Let us disregard them, just for now, and try to make sense of the catenary equation.

To our arch equation, in simplified form becomes

$$y = e^x + e^{-x}$$

Now this equation may still be daunting to some, so I will try to break it down, explain each piece, and then put it back together.

Exponential Growth

Let us start with that letter e. It is not like the constant a in our earlier equation, that stands for any constant value. It is a very special number, and has a particular value.

There are different ways to arrive at the value of this number e, but perhaps the most intuitive starts with the familiar idea of compound interest, like that given by a bank.

If you invest P dollars at an interest rate n, you would have an amount y after x years,

$$y = P(1+n)^x \text{ dollars}$$

For example, $500 invested at 6.5% for 8 years gives

$$y = 500(1 + 0.065)^8 = \$827.50$$

For this example, the interest was compounded *once a year*.

Now suppose we compound interest m times per year. We would now have mx interest periods, but for each the interest rate would be n/m.

$$y = P\left(1 + \frac{n}{m}\right)^{mx}$$

Repeating our example with interest compounded monthly gives

$$y = 500\left(1 + \frac{0.065}{12}\right)^{12(8)} = \$893.83$$

or another $12.33.

Fine, but what does this have to do with the number e? We will see if we now ask, *what if the interest were not computed in discrete steps, but continuously?* Suppose m were not 12 or 365 or 1000, but *infinite*.

Let us make the substitution

$$k = \frac{m}{n}$$

so that our equation for compound interest becomes

$$y = P\left(1 + \frac{1}{k}\right)^{knx},$$

which can be written

$$y = P\left[\left(1 + \frac{1}{k}\right)^k\right]^{nx}.$$

Now as m gets very large, and k also gets very large, what happens to the quantity inside the brackets? Let us calculate some values.

k	$1+\dfrac{1}{k}$	$\left(1+\dfrac{1}{k}\right)^{k}$
1	2.0000000000	2.00000
10	1.1000000000	2.59374
100	1.0100000000	2.70481
1000	1.0010000000	2.71692
10000	1.0001000000	2.71815
100000	1.0000100000	2.71827
1000000	1.0000010000	2.71828

We get the surprising result that as m becomes infinite, the value of the expression in the brackets settles down to a specific value, about 2.7183. *That value is called e.*

Now replacing $(1+1/k)^{k}$ by e in our formula for compound interest gives the formula for continuous growth or exponential growth:

$$y = Pe^{nx} .$$

Does that look familiar? Its one of the expressions in our arch equation, with $P = 1$ and $n = 1$.

Populations offer an example of exponential growth; the greater the population the faster it grows, or

The rate of growth is proportional to the amount.

Fig. 5 is a graph of exponential growth, with $P = 1$ and $n = 1$.

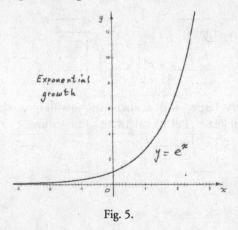

Fig. 5.

Exponential Decay

Now lets look at the other term in the catenary equation, e^{-x}. The minus sign in the exponent is all that is needed to change the equation from describing exponential growth into one that describes exponential decay (fig. 6).

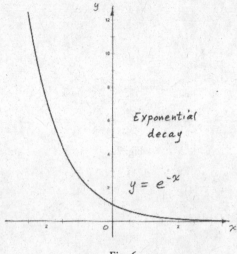

Fig. 6

As an example, take a hot cup of coffee. It is the temperature *difference* between the coffee and the room air that drives heat out of the coffee. As the coffee cools, the temperature difference decreases, so the *rate* of temperature drop decreases. As with population growth,

The rate of change of temperature is proportional to the temperature T.

In calculus notation,

$$\frac{dT}{dt} = kT$$

The Hanging Chain

Now let us plot the growth and decay curves on the same axes, and also graph *their sum.* Note that the sum of the growth and decay curves combine to form the *catenary*.

The seventeenth-century Dutch mathematician Christian Huygens named the curve *catenarius*, after the Latin word for *chain*. The reason for that name is that this curve describes the shape of a uniform chain or flexible rope hanging from two points.

What does a hanging chain have to do with exponential growth and decay?

Let us look at the forces on a small section of cable. Fig. 7 shows the tensions T_U and T resolved into vertical and horizontal components.

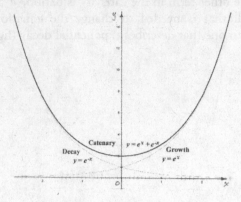

Fig. 7.

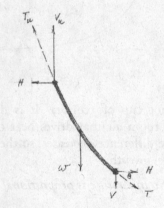

Fig. 8.

We note the following:

- The two horizontal forces H must be equal and opposite, as there are no other horizontal forces acting on this section of cable.

- The vertical force V is equal to half the weight of the cable below that point, the other half being supported on the right-hand side.

- As we go lower along the cable, the vertical force V decreases a constant amount for each foot of cable.

- As the vertical force decreases, the angle θ of the resultant decreases, as the horizontal force H is constant.

Thus the slope of the curve decreases as we move along it, so the rate of change in vertical distance is proportional to the distance along the curve. But for steeper parts of the curve, the vertical displacement is approximately equal to the distance along the curve (fig. 8).

The rate of change of vertical distance is approximately proportional to the vertical distance.

Comparing this with temperature drop in our coffee cup, we saw that

The rate of change of temperature was proportional to the temperature T of the coffee.

In calculus notation,

For drop in temperature: For drop in height:

$$\frac{dT}{dt} = kT \qquad\qquad \frac{dy}{dx} \cong ky$$

So the descending portion of the curve, at least where it is steep, behaves something like exponential decay. In a similar way we can show that the rising portion of the curve exhibits some characteristics of exponential growth.

Why use a catenary for the St.Louis Arch?

With the hanging chain, the weight of the links gets resolved into tensions that always act along the curve, and never at right angles to it. There is pure tension throughout. Similarly for the catenary arch, the weight of the arch acts along the centerline of the arch, and there are no shear forces perpendicular to the centerline.

Why is a hanging chain described by the catenary equation?

We glibly stated that a hanging chain follows the catenary curve, but why? The proof of that statement is a standard problem in many mechanics or differential equation textbooks. This one is from Murray Spiegel's *Applied Differential Equations,* Prentice-Hall, 1958.

We start with a bit of chain or rope with one end at the low point C of the curve. It is acted on by three forces, H, W, and T (fig. 9).

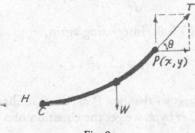

Fig. 9.

From the equations of equilibrium,

$$T \sin \theta = W$$
$$T \cos \theta = H$$

Dividing gives

$$\frac{\sin \theta}{\cos \theta} = \tan \theta = \frac{W}{H} = \text{slope at } P = \frac{dy}{dx}$$

Taking the derivative, noting that H is a constant

$$\frac{d^2 y}{dx^2} = \frac{1}{H} \frac{dW}{dx} \qquad (1)$$

Here $\frac{dW}{dx}$ is the change in load per unit *horizontal* distance. We will do two derivations: one for the suspension bridge and another for the catenary.

For a ***suspension bridge*** for which the weight of the cable is negligible compared to the weight of the roadway, this change in load is constant, and equal to the weight per foot of the roadway, which I'll call w (fig. 10). Then Eq. (1) becomes

$$\frac{d^2 y}{dx^2} = \frac{w}{H}$$

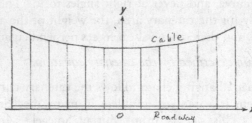

Fig. 10.

Integrating gives

$$\frac{dy}{dx} = \frac{w}{H} x + C_1$$

But $dy/dx = 0$ at $x = 0$, so $C1 = 0$. Integrating again,

$$y = \frac{w}{2H} x^2 + C_2$$

We can now shift the y axis so that $y = 0$ at $x = 0$. Then C^2 will be zero. So, replacing the constant $w/2H$ by k, we get the equation of a *parabola*

$$y = kx^2$$

Turning now to the catenary, the load per unit distance *along the curve, dW/ds,* is constant, and has the value *w,* but the load per unit *horizontal* distance is not constant. For the steeper parts of the curve, the load per unit horizontal distance is greater than for less steep portions.

So we must write an expression for *dW/dx* as a function of *x.*

$$\frac{dW}{ds} = w = \left(\frac{dW}{dx}\right)\left(\frac{dx}{ds}\right)$$

So,

$$\frac{dW}{dx} = w\frac{ds}{dx} \quad (2)$$

Now let us look at a section of the curve so small that it can be considered straight (fig. 11).

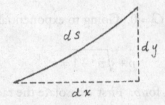

Fig. 11.

By the Pythagorean theorem,

$$\left(\frac{ds}{dx}\right)^2 = \left(\frac{dy}{dx}\right)^2 + \left(\frac{dx}{dx}\right)^2$$

from which, with *dx/dx* = 1,

$$\frac{ds}{dx} = \sqrt{\left(\frac{dy}{dx}\right)^2}$$

So from (2),

$$\frac{dW}{dx} = w\sqrt{1 + \left(\frac{dy}{dx}\right)^2}$$

Substituting into (1) gives

$$\frac{d^2y}{dx^2} = \frac{w}{H}\sqrt{1 + \left(\frac{dy}{dx}\right)^2} \quad (3)$$

Here we have a second-order differential equation which we now solve for y. Let

$$\frac{dp}{dx} = p \text{ and } \frac{H}{w} = C$$

So (3) becomes

$$\frac{dp}{dx} = \frac{1}{C}\sqrt{1 + p^2}$$

Separating variables,

$$\frac{c \, dp}{\sqrt{1 + p^2}} = dx$$

Integrating, using a rule from a table of integrals,

$$\ln\left(p + \sqrt{p^2 + 1}\right) = \frac{x}{C} + C_3$$

But $p = \frac{dy}{dx} = 0$ at $x = 0$, so $C_3 = 0$. Going to exponential form,

$$p + \sqrt{p^2 + 1} = e^{x/C}$$

We now solve this equation for p. First we isolate the radical:

$$\sqrt{p^2 + 1} = e^{x/C} - p$$

Squaring both sides,

$$p^2 + 1 = e^{2x/C} - 2pe^{x/C} + p^2$$

$$2pe^{x/C} = e^{2x/C} - 1$$

$$p = \frac{1}{2}\left(\frac{e^{2x/C}}{e^{x/C}} - e^{-x/C}\right)$$

$$p = \frac{1}{2}\left(e^{x/C} - e^{-x/C}\right) = \frac{dy}{dx}$$

Replacing p by dy/dx and integrating again,

$$y = \frac{C}{2}\int e^{x/C}\left(\frac{dx}{C}\right) - \frac{-C}{2}\int e^{-x/C}\left(\frac{-dx}{C}\right)$$

$$y = \frac{C}{2}\left(e^{x/C} + e^{-x/C}\right) + C_4$$

We can shift the y axis so that $C_4 = 0$. Our equation then becomes

$$y = C\left(\frac{e^{x/C} + e^{-x/C}}{2}\right),$$

the equation of the *catenary*.

Catenary and parabola compared

If we draw a parabola that has the same height and width as the St. Louis arch would have the equation

$$y = 0.00699x^2$$

Fig. 12 shows this equation plotted together with the actual arch equation.

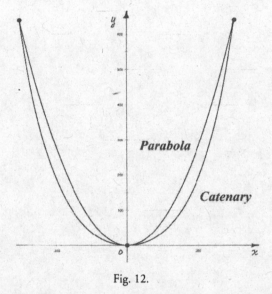

Fig. 12.

Notice that the catenary is flatter than the parabola. This is due to the weight distribution for each.

The suspension bridge has the same weight per horizontal foot, while the catenary does not. There is relatively more weight away from the centerline, where the curve is steeper. Thus there is more chain in each one-foot interval towards the ends than at the center. This tends to make the catenary lower away from the centerline.

About the author

Paul A. Calter is Professor Emeritus of Mathematics at Vermont Technical College. He has interests in both the fields of mathematics and art. He received his B.S. from Cooper Union and his M.S. from Columbia University, both in engineering, and his Masters of Fine Arts Degree at Vermont College of Norwich University. Calter has taught mathematics for over twenty-five years and is the author of ten mathematics textbooks and a mystery novel. He has been an active painter and sculptor since 1968, has participated in dozens of art shows, and has permanent outdoor sculptures at a number of locations in Vermont. Calter developed the Mathematics Across the Curriculum course "Geometry in Art & Architecture" and has taught it at Dartmouth and Vermont Technical College, as well as giving workshops and lectures on the subject. He presented a paper on the survey of a doorway by Michelangelo in the Laurentian Library in Florence at the Nexus 2000 conference on architecture and mathematics. Calter developed a trigonometric method for non-contact measurements of facades and presented his method at the first Nexus conference in 1996. His book, *Squaring the Circle: Geometry in Art and Architecture*, is due out in January 2007.

Don Hanlon

School Of Architecture and
Urban Planning
University of Wisconsin –
Milwaukee
PO Box 413, Milwaukee,
WI 53201-0413 USA
dhanlon@uwm.edu

keywords: architecture,
culture, construction
technology, arch

Research

Arches and Culture

Abstract. Technological innovation is the driving force of our civilization. Therefore, we assume all other civilizations would exploit a technological advantage to the same degree that we would. We forget, however, that technology is an aspect of culture, and as with any other aspect of culture, it may be more or less important to any given civilization. The history of the arch is an interesting case in point. The arch is a structural device in architecture that has distinct advantages over post-and-beam construction. People have known how to build the arch and how to use it since the third millennium B.C., but did not use it because its form and meaning did not fit with other dimensions of their respective cultures.

Introduction

In the last three hundred years, the mainstream of western architectural history and criticism depended on two methods of measuring value. The first, predominantly of the eighteenth and nineteenth centuries, was the Romantic appreciation of art. Its project was to categorize historical styles based on European aesthetic judgment. Aesthetic value, or Beauty, resided in that which was virtuous or truthful from a European perspective, so the architectural debates of the nineteenth century centered, to a great extent, on religious or moral virtues. For example, Ruskin contrasted what he claimed were the authentic European Christian virtues of the Gothic style against the decadent orientalism of Classical architecture. The second measure of architectural value emerged from industrialism in the late nineteenth and early twentieth centuries. Technological rationalism argued that Beauty resides in that which is the most direct, objective, and logical solution to a physical problem, using the most up-to-date materials and technologies available.

Art appreciation and technological rationalism have continued to be the dominant ways by which we judge the merits of architectural form in our own culture and we continue to project the same set of values onto those cultures which preceded us. However, these products of specific social, political, and economic conditions of early Modernism prevent us from understanding the motivations of architectural designers who built before the modern era. Ancient people did not create art and architecture for the aesthetic pleasure of modern people. To judge the merits of ancient architecture solely on aesthetic grounds is to presume that ancient architects did so as well and, moreover, shared our system of values. Likewise, technological rationalism presumes that ancient people judged the success or beauty of their architecture by values that arose from an equivalent of industrialism. If the architects who built the pyramid of Cheops in the third millennium BC had access to steel and concrete, would they have built the St.

Nexus Network Journal 8 (2006) 67-72
1590-5896/06/020067-6 DOI 10.1007/s00004-006-0018-6

Louis arch instead? Would the Greek architects, Ictinus and Calicrates, have built the Sears Tower instead of the Parthenon?

The role of the arch – and its derivative, the vault – in architectural history reveals a third way of understanding how values shape architecture. Though the arch can be understood aesthetically as a sculptural form and technologically as a structural form, it is first and foremost a *cultural* form because ancient people distinguished architecture from mere building not on the basis of aesthetics or technological ingenuity, but on the basis of *cultural meaning*. To illustrate this assertion, let us examine two architectural traditions, the Egyptian and the Greek. There is extensive evidence to show that both the Egyptians and the Greeks understood the arch as a structural device and had the technological sophistication to use it to full advantage, yet they avoided it or relegated it to minor, inconsequential applications because it did not fit into the web of meanings that dominated their respective cultures.

The arch in Egypt

In the earliest archaeological record of Egypt, we find depictions of small structures with curved roofs. The roofs were not arches or vaults, however. The walls and roofs of these houses were composed of lightweight bent wooden frames, or bundles of reeds, covered with woven mats or textiles. This was a simple device that exploited materials easily at hand. As the form of the common house became associated with the ruler and eventually with divine beings, it became a stylized pavilion. Since Egyptian civilization was profoundly conservative, the architectural forms that emerged at its inception persisted throughout its entire history, unchanged in form but varying in material. The small wooden house became the model for subterranean "houses" for the dead built of sun-dried brick. By the beginning of the Third Dynasty (ca. 2686 BC), the Egyptians were masters of brick construction, including the use of arches and vaults. However, their only application of this technology was in burial chambers below ground where brick was more durable than wood and where the lateral thrust of an arch or a vault was easily absorbed by the surrounding compacted earth. For example, the tombs at Reqâqnah and Bêt Khallâf are evidence of the Egyptians' understanding of the mechanics of the arch [Garstang 1904, 22]. These were true arches and vaults, composed either of standard bricks with a wedge-shaped infill of mud mortar or of specially shaped bricks that acted as voussoirs. Because of the Egyptian obsession with the continuity of form from life through death, these tombs were subterranean replicas in brick of the terrestrial wooden house, complete with curved roof [Smith 1968, 55].

The mortuary complex of Zoser at Saqqara was the first monumental stone architecture in history. Within the complex, a single example illustrates that the Egyptians understood the structural advantage of an arch at a very early date and knew how to construct it in stone. It is the small relieving arch, with cut stone voussoirs, in the tomb of Hebset-Neferkara. However, this too is below ground. The Egyptians appear to have had no interest in using the structural advantages of the arch above ground. Nowhere do we find the light rhythmic effects produced by arcaded walls. Likewise, their use of vaults above ground was characteristically conservative, limited in almost all cases to small utilitarian structures, typified by the long, narrow storage cells in the Ramesseum (Nineteenth Dynasty) [Smith 1968, 134]. The technique of placing bricks to create these tunnel vaults is called *laminated vaulting*, by which masons laid brick courses at an angle against one end wall, each course bearing on its predecessor. This technique does not require timber centering and is still used, three millennia later, in rural Egypt. One of the rare uses of what appears to be a vault in a sacred space is in the Sanctuary of Osiris at Abydos (also of the Nineteenth Dynasty). It is not a true vault, however, but a corbelled one, in which horizontal courses of stone were carved into a curve [Smith 1968, 169]. The intention here was not to create a new kind of architectural space, but to create an interior space, in stone, reminiscent of an archaic house with a curved roof composed of bent reeds and matting.

Also at Saqqara, the elegantly curved roofs of the shrines facing the Heb-Set Court give the impression of vaults, but they are not. Their non-structural curved stone skins merely covered solid brick cores. Once again, the intention was to perpetuate an ancient tradition of architectural form, the sacred house, which conveyed a fundamental meaning in Egyptian culture: the fixed linkage of the present with the primordial past [Lauer 1976, 135]. These examples illustrate that brick and stone came into use in Egypt not to provide a new aesthetic, or to be explored and celebrated as technological innovations, but rather to preserve immutable architectural stereotypes in buildings of increasing scale and durability. The technological sophistication that the Egyptians brought to bear upon the construction of the pyramids – the cutting, transport, and setting of stones weighing many tons with great precision – clearly shows a level of ingenuity and invention that could have been devoted to solving the problems of lateral thrust produced by arches and vaults above ground. The same basic engineering that was used to construct simple round arches could have been turned to producing large, unified interior spaces spanned by vaults. To the contrary, the Egyptians showed no interest in this line of thought because these forms and spaces had no cultural relevance. Instead, Egyptian monumental architecture – such as the temples at Karnak, Deir el-Bahari, and Luxor – relied exclusively on a trabeated structure of columns and beams that replicated an ancient tradition of diminutive domestic structures in wood at a colossal scale in limestone.

The arch in Greece

The Greeks were among the greatest practitioners of masonry construction in the entire history of architecture. For example, their superb technique for constructing walls with cyclopean polygonal stones was exceeded only by the masons of the Incan Empire. However, the Greeks' craftsmanship was not matched by structural experimentation. Their use of the arch and the vault was highly conservative and examples before the Roman era are rare. This cannot be explained by ignorance, since they had extensive and prolonged contacts with Oriental cultures that used the arch extensively. The arch and the vault had been used in Western Asia for many centuries prior to the Greeks arrival, but these examples were composed of brick, a material in which the Greeks showed little interest [Lawrence 1983, 170]. The Greeks used the arch or vault of stone only in situations where the lateral thrust was easily counteracted by massive walls, as in fortifications, or by the earth in subterranean chambers. However, the Greeks' reticence to use the arch and the vault cannot be explained by a lack of technical knowledge or skill. The primary factor was cultural. Similar to the Egyptians, the Greeks established a formal language for their architecture at a very early date and never deviated from its rules. Their architecture was of a trabeated type, composed of columns and beams, originally of timber and later translated into stone. Every element and detail of the Classical canon of architecture that the Greeks used in their monumental temples constructed of stone, such as the Parthenon, has a direct correlation to a timber component in their archaic domestic buildings. It was for cultural reasons that the Greeks chose not to use any structural system other than a trabeated system in their major civic and religious buildings. Their urban form, the *polis,* was the basis of Greek identity and defined a group as a distinct racial entity with an unbroken historical and mythic connection to the origins of their civilization. Likewise, their ancestral architectural form was the *megaron*, a small dwelling composed of a more or less cubic interior volume, with a colonnaded porch, both protected by a single gable roof. This was the archetypal building from which the Greeks derived the designs for all of their civic and religious structures. Even their grandest and most elaborate buildings were variations on the theme of the simple *megaron*.

The Greek temple was foremost a symbol of the sanctuary, or shelter for a cult deity, a deity that protected and perpetuated the identity of a people. As a symbol for an immutable principle, its form could not change. Therefore, the entire architectural project of the Greek canon, extending over a millennium, was to only make incremental refinements to an established model. Furthermore, monumental Greek architecture was to be contemplated as sculpture in brilliant sunlight and in relation to its surrounding landscape; it was not dedicated to enclosing interior space. Therefore, the Greeks lacked an incentive to experiment with structural systems such as the arch, vault or dome that would have expanded or elaborated interior volumes.

Conclusions

The two Modernist methods of measuring value in architectural history – art appreciation and technological rationalism – do not help us to understand the ways in which people in preceding cultures thought. The archaeological plunder that we consider "art" had no artistic significance, in the Modern sense, to the people who created it. Whether it was a small cult figure or a monumental building, it was a ritual object whose sole purpose was to preserve cultural meanings. These artifacts comprised a cultural landscape that fused myth, history, language, and social relations in a single form. Our contemporary aesthetic values can at best suggest a superficial parallel experience.

The application of technological rationalism reveals an even greater disassociation of ideas. The ideology of modern culture is technological innovation. Technology is the single greatest determinant of our well-being and of our self-image, both as individuals and as a civilization, because it shapes a set of values that determines for us the relevance of knowledge and experience. In our view, the advance of technology is inexorable and irresistible, so we assume that every technological advance must be exploited, regardless of the consequences. It is our mythology, the overarching religion of our age. In our daily lives, technology drives time, deconstructs space and creates it anew in strange and compelling forms. Anyone who resists is considered naïve, anachronistically Romantic, or off their medication. After all, is not the advance of technology a fact of life? It is inconceivable that entire civilizations would not only fail to exploit a technology that offers distinct advantages, but even worse, actively avoid it. However, the reason we feel this way is because we have forgotten that technology is a cultural phenomenon, and as such, it is susceptible to the forces of other dimensions of culture, such as language and myth. Each civilization uses a set of values to determine the relative importance of the many kinds of experiences that comprise their culture. We have placed technology at the pinnacle of our hierarchy of relevance, but other cultures have not. If we judge those civilizations by our set of values, we will never know them, and will probably simply dismiss them as ignorant and primitive. The civilizations of the Egyptians and the Greeks rejected an architectural technology that had obvious practical advantages because it failed to reinforce or elaborate the underlying forms of their respective cultures. They made a conscious choice to pursue meaning in the physical and mythic worlds as an integrated whole within the limits of familiar cultural patterns.

Bibliography

GARSTANG, John. 1904. *Tombs of the Third Egyptian Dynasty at Reqâqnah and Bêt Khallâf* (Westminster: Archibald Constable).

SMITH, Earl Baldwin. 1968. *Egyptian Architecture as Cultural Expression* (Watkins Glen: American Life Foundation).

LAUER, Jean-Philippe. 1976. *Saqqara – The Royal Cemetary of Memphis* (London: Thames & Hudson).

LAWRENCE, A.W. 1983. *Greek Architecture*, 5 ed. (New Haven: Yale University Press).

About the author

Don Hanlon is an architect and Professor, focusing on design, history, and theory. His research and writing has concentrated on cultural aspects of architecture and urban design, primarily in Islamic societies in Egypt, India, and Central Asia. His interest in cultural forces at work in the history of architecture has also included papers and lectures concerning the uses of architectural symbolism by Native Americans in their resistance to colonialism, and the origins and morphology of the American gas station.

Michael Serra

163 Mendosa Ave.
San Francisco, CA 94116
USA
mserra@earthlink.com

keywords: arch construction,
vouissoir, Interior Angle Sum
Conjecture, Exterior Angle Sum
Conjecture, triangles, trapezoids

Research

Solving Ertha Diggs's Ancient Stone Arch Mystery

Abstract. According to legend, when the Romans made an arch, they would make the architect stand under it while the wooden support was removed. That was one way to be sure that architects carefully designed arches that wouldn't fall! Educator Michael Serra led AAAS symposium participants in a surprising and fun hands-on arch construction project using familiar objects—Chinese take-out cartons—in an unfamiliar way: "these are stone voussoirs from an ancient miniature bridge uncovered by my friend, archaeologist Ertha Diggs. She has asked us to determine the number of stones in the original bridge." This makes it possible to understand both arch mechanics and the mathematics behind the arch through actually constructing them.

Introduction

These lessons took place in a geometry class in which students work in cooperative groups of four to discover the properties of geometry. They are familiar with the basic tools of geometry: compass and straightedge, patty papers, and The Geometer's Sketchpad. They have recently discovered the Interior Angle Sum Conjecture, the Exterior Angle Sum Conjecture, and properties of isosceles triangles and trapezoids (fig. 1).

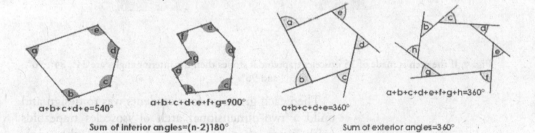

$a+b+c+d+e=540°$

$a+b+c+d+e+f+g=900°$

$a+b+c+d+e=360°$

$a+b+c+d+e+f+g+h=360°$

Sum of interior angles=(n-2)180°

Sum of exterior angles=360°

Fig. 1.

Lesson: Part 1

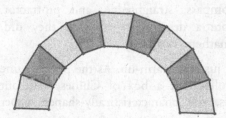

I began the lesson by sharing photos and transparencies of Roman and Chinese arches while discussing their history and development. Whether true or not, my students especially enjoyed the tale about the Roman architect for an arch. During class discussion we arrived at the

1590-5896/06/020073-6 DOI 10.1007/s00004-006-0019-5

geometric characteristics of an arch. We agreed that the arch is half of a regular polygon. We conjectured that half the number of sides of the regular polygon must be an odd number (in order to have a keystone). For example, half of a regular 18-gon gives us an arch with 9 stones, 8 voussoirs and a keystone (fig. 2).

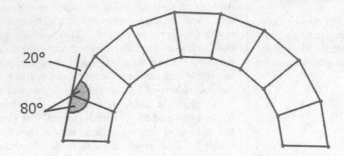

Fig. 3. If the arch is made of 9 isosceles trapezoidal stones then the interior angles are 80°, 80°, 100°, and 100°.

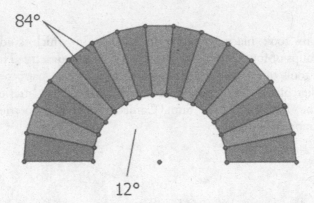

Fig. 4. If the arch is made of 15 isosceles trapezoidal stones then the interior angles are 84°, 84°, 96°, and 96°.

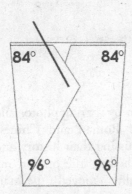

Then each group of four students was to design and build a two-dimensional arch of isosceles trapezoids (with more than 9 stones). I required the angle measures of the isosceles trapezoids to be positive integers. The students then discussed, planned, designed, and constructed their two-dimensional arches with compass, straightedge, and protractor. Next they wrote a description of what they did, describing the mathematics used.

But that was just the warm-up. As the period came to a close I pulled out a box of Chinese take-out cartons, that is, the characteristically-shaped paper boxes used by Chinese restaurants (fig. 5). I told them, in mock scientific seriousness, that "these are stone voussoirs from an ancient miniature bridge uncovered by my friend, archaeologist Ertha Diggs. She has asked us to determine the number of stones in the original bridge." I gave one to each group, the bell rang and class was dismissed.

Lesson: Part 2

When the students came in to class the next day they began discussing the problem posed yesterday in their groups. I interrupted the group discussions to announce that when a group determines how many pieces in the original (Chinese take-out carton) bridge they are to write up an explanation and then call me over. When they call me over and explain their reasoning, I then give them the additional cartons they think they need to build a replica of the bridge. By the end of the period, when each group has built their arch, we bring them all together and assemble them into a vault! Of course these teenagers cannot resist crawling through the vault. The objective was for students to review and apply the properties of isosceles triangles, trapezoids, regular polygons, and of interior and exterior angle sums. They were to practice communicating mathematically and modeling in two and three dimensions. It is a fun two-day activity of hands-on mathematics and problem solving.

Follow-up

During the previous summer there was a lot of remodeling at our school site. So we had a large number of cardboard boxes stacked in piles all over the school building. When I see a free resource lying around I am compelled to make use of it in some way. So I created a lesson on nets for solids. On Day one, I asked my students to explore all the possible non-congruent nets for a particular isosceles trapezoidal prism (fig. 6).

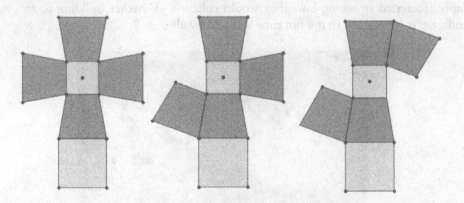

Fig. 6. Three isosceles trapezoidal prisms are shown above. Can you find the others?

When completed we selected the one to best fit on one of the folded flat cardboard boxes. Once we had the largest possible net designed on the cardboard we passed it around and each group traced it onto their cardboard. The next day they cut and assembled each net into the isosceles trapezoidal stone ready for assembly into an arch. Would the arch be large enough for us to walk under? Having recently completed the trigonometry chapter I was determined to get them to apply their new trig skills. The task of each group was to measure their trapezoidal stone and calculate the span and rise of the arch that was going to be created by these voussoirs (fig. 7).

Fig. 7.

If a (length of large base) = 26 cm

and b (length of small base = 22 cm,

then

Span = $2r = b/\sin 6° = 22/\sin 6° \approx 210.5$ cm

Rise = $h = .5b/\tan 6° = 11/\tan 6° \approx 104.7$ cm

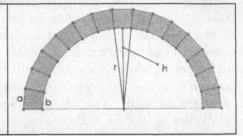

By the second and third geometry classes of the day we had enough voussoirs to complete the arch. It was agreed that I was the chief architect of the arch so I was required to remain beneath the arch as all supporting hands were removed. I managed to survive the last stage of construction. Once again Geometry and Architecture blended beautifully into a few days of fun applications for my geometry students and myself. At the St. Louis Arch symposium (during the AAAS conference) in February of 2006 I had the pleasure of watching a group of professional architects and professors of architecture get on their hands and knees and play with my Chinese take-out carton arches. Whereas high school students and teachers that I have worked with in the past are fascinated with the building of arches and vaults, my new students, these architectural professionals, seemed much more interested in seeing how they would collapse! Whether building or forcing collapse, it was agreed that a fun time was had by all (figs 8-12).

Fig. 8. AAAS symposium participants studying the "relics" found by Ertha Diggs

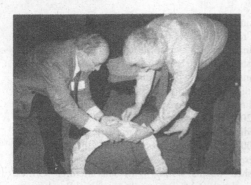

Figs. 9 and 10. Studying the constructed arch

Fig. 11. Manipulation of the arch to see the various deformations that lead to collapse

Fig. 12. AAAS symposium speakers Matthys Levy, Santiago Huerta (holding arch), Kim Williams, Paul Calter (behind arch), John Ochsendorf and Michael Serra

About the author

Since the 1990 publication of the first edition of *Discovering Geometry: An Investigative Approach* (*DG*) (San Francisco: Key Curriculum Press), Michael Serra has been a popular classroom teacher, speaker, and teacher trainer at Geometry Institutes. When not teaching, he is either writing new material or traveling the country giving workshops to districts that have already adopted or are thinking of adopting *DG*. He also gives presentations at a number of National Council of Teachers of Mathematics (NCTM) regional conferences, or state mathematics conference around the country. Internationally, Michael has consulted for mathematics teachers in Costa Rica, South Africa, Australia, and international schools in China. In 2002 the third edition, *Discovering Geometry: An Investigative Approach* was released. The fourth edition of *Discovering Geometry: An Investigative Approach* is scheduled for release in January of 2007. Other publications include the very popular supplementary geometry book, *Patty Paper Geometry*, and the set of five classroom starter workbooks *Mathercise*. Key Curriculum Press has just published *What's Wrong With This Picture? Critical Thinking Exercises in Geometry*. His next project is a book of Mathematical Games and Puzzles.

Elena Marchetti
Luisa Rossi Costa

Department of Mathematics
Milan Polytechnic
Piazza Leonardo da Vinci, 32
20133 Milano, Italy
elena.marchetti@polimi.it
luisa.rossi@polimi.it

keywords: mathematics and
architecture, geometry, regular
curves, quadric surfaces

Research

Mathematical Elements in Historic and Contemporary Architecture

Abstract. Starting from the idea that Mathematics plays an important role in planning any aesthetically attractive and functional construction, this work focuses on curves and surfaces easily recognisable in buildings. Many contemporary examples, but also some intriguing forms connected by classical geometrical questions are illustrated. Nowadays as well in the past, architects often give a splendid interpretation of the beauty of Mathematics; at the same time they introduce modern aspects of this important subject, related to the social and environmental field.

...The circle is used to represent perfection, the dome of the heavens. The square represents the heart, the four elements and the rational intellect...(Richard Meier).[1]

Introduction

The surfaces externally delimiting significant buildings and the forms characterizing their interiors are usually very elegant and well proportioned.

Some observers consider these shapes only from an artistic point of view and they appreciate principally the aesthetic aspect; others, because of their scientific-technical background, try to understand the structural and mathematical properties as well. These last conscious that the pleasant harmony of an architectural structure comes also from its intrinsic geometrical qualities and from the structural and decorative materials employed.

Therefore as mathematicians interested in architecture and arts, we would like to confirm the fact that the shape's harmony comes not only from the artist's creativity but it is strictly connected with mathematical rules. The more you know mathematical tools the more you understand and appreciate the form's beauty.

In this paper we intend to focus on curves and surfaces often involved in façades, domes and vaults of buildings. The aim is to illustrate those forms in mathematical way, to underline their configuration and some essential particulars not always evident at first glance.

We take this occasion also to notice that the contemporary architects consider important technical aspects as well, to make buildings more functional, to minimize energy costs or to facilitate insertion in the social, environmental context. Consequently mathematics enters deeply in planning (mathematical modelling) in order to realize a better performance.

In the first section we introduce the mathematical elements used to represent curves and surfaces; in the second section we describe and generate different forms, as recognisable in classical monuments as in recent buildings, among them skyscrapers [Höweler 2004]. In the third section we present an interesting contemporary use of portions of cylinders and spherical surfaces: the elementary geometric parts are involved again but they produce a totally different effect in the buildings, thanks to the creativity of the architects.

Mathematical tools

We introduce the essential mathematical elements to describe curves and surfaces; among different possibilities we select the parametric form, one of the most adaptable and efficient way to represent the geometrical loci as graphically as well as analytically. More in detail, in the 3D Cartesian orthogonal space $Oxyz$, a

regular curve is represented by a vector $\mathbf{v}(t) = \begin{bmatrix} x(t) \\ y(t) \\ z(t) \end{bmatrix}$, whose components depend

on the _real parameter_ $t \in I = [a,b] \subset R$, $\mathbf{v}(t) \in C^1(I)^2$ and $\|\mathbf{v}'(t)\| = \sqrt{x'(t)^2 + y'(t)^2 + z'(t)^2} \neq 0$, that is, the usual _conditions of regularity_ are satisfied.

A _regular surface_ is equally represented by a vector $\mathbf{w}(t,u) = \begin{bmatrix} x(t,u) \\ y(t,u) \\ z(t,u) \end{bmatrix}$, depending

on _two real parameters_ $(t,u) \in A = ([a,b] \times [c,d]) \subset R^2$, $\mathbf{w}(t,u) \in C^1(A)$ and $\|\mathbf{w}_t \wedge \mathbf{w}_u\| \neq 0$.[3]

In both cases, for curves as well as for surfaces, the parameterisation is not unique.

Hereby we mention _basic regular curves_ and choose a standard parameterisation:

Straight line:	$\mathbf{v}(t) = \begin{bmatrix} x_0 + \alpha t \\ y_0 + \beta t \\ z_0 + \gamma t \end{bmatrix}, t \in R$;	the point $P_0 = (x_0, y_0, z_0)$ belongs to the line having the direction of the vector $\begin{bmatrix} \alpha \\ \beta \\ \gamma \end{bmatrix} \neq 0$.
Ellipse:	$\mathbf{v}(t) = \begin{bmatrix} r_1 \cos t \\ r_2 \sin t \\ 0 \end{bmatrix}, t \in [0, 2\pi]$	where $r_1, r_2 \in R^+$ (that is positive real numbers) are the semi-axes of the ellipse centred in O and belonging to the xy plane; the ellipse becomes a circle if $r_1 = r_2$.

Parabola:	$$\mathbf{v}(t) = \begin{bmatrix} t \\ at^2 + bt + c \\ 0 \end{bmatrix}, t \in R$$;	and the coefficients $a, b, c \in R$; the parabola, belonging to the xy plane, has symmetry axis parallel to y-axis.
Spiral:	$$\mathbf{v}(t) = \begin{bmatrix} f(t)\cos t \\ f(t)\sin t \\ 0 \end{bmatrix}, t \in R$$	where $f(t) \in C^1(R)$ is a nonnegative, monotone function; the choice $f(t) = at, a \in R^+$, gives *Archimedean spirals*, the choice $f(t) = e^{at}, a \in R$, gives *logarithmic spirals*.

We also introduce *basic regular quadric surfaces* such as:

Ellipsoid:	$$\mathbf{w}(t,u) = \begin{bmatrix} a\sin t\cos u \\ b\sin t\sin u \\ c\cos t \end{bmatrix}$$;	$(r,u) \in ([0,\pi] \times [0,2\pi])$ and $a,b,c \in R^+$; the coefficients a, b, c give the semi-axes of the ellipsoid, centred in O. The ellipsoid becomes a sphere if $a = b = c$.
Circular paraboloid:	$$\mathbf{w}(t,u) = \begin{bmatrix} u\cos t \\ u\sin t \\ cu^2 \end{bmatrix},$$	$(r,u) \in ([0,2\pi] \times R^+)$ and $c \in R$; the z-axis is the symmetry rotation axis and the vertex is placed in O.
Cylinder:	$$\mathbf{w}(t,u) = \begin{bmatrix} a\cos t \\ b\sin t \\ u \end{bmatrix}$$;	$(r,u) \in ([0,2\pi] \times R)$ and $a,b \in R^+$; the directrix is an ellipse and the generatrices are parallel to z-axis. Different cylinders can be described taking a regular simple curve as directrix.
Cone:	$$\mathbf{w}(t,u) = \begin{bmatrix} au\cos t \\ bu\sin t \\ u \end{bmatrix}$$	$(r,u) \in ([0,2\pi] \times R)$ and $a,b \in R^+$; the vertex is placed in O and the sections orthogonal to the z-axis are ellipses (directrices). Other conic surfaces come from the different choice of the directrix.
Hyperboloid of one sheet:	$$\mathbf{w}(t,u) = \begin{bmatrix} aChu\cos t \\ bChu\sin t \\ cShu \end{bmatrix}$$	$(r,u) \in ([0,2\pi] \times R)$ and $a,b,c \in R$; sections orthogonal to the z-axis are ellipses.
Hyperbolic paraboloid:	$$\mathbf{w}(t,u) = \begin{bmatrix} u \\ t \\ a^2t^2 - b^2u^2 \end{bmatrix}$$	$(t,u) \in (R \times R)$ and $a,b \in R$; sections orthogonal to the z-axis are hyperbolas; sections orthogonal to the x- or y-axis are parabolas.

Fig. 1. a) Lyon Opera Theatre by J.Nouvel (photo by L. Rossi); b) Cooling tower in Milan (photo by E. Marchetti); c) Guggenheim Museum, Bilbao by F. Gehry (photo by L. Rossi); d)The Willis building, Ipswich, by N. Foster (photo by L. Rossi)

We recall that cylinder, cone, hyperboloid of one sheet and hyperbolic paraboloid are *ruled surfaces*, that is through each point of them passes at least one straight line that lies entirely in the surface.

Frequently *ruled surfaces* contain two families of rulings, such as the hyperboloid of one sheet (*doubly ruled*).

The equations given above can be combined and adapted to the geometry of many shapes. Cylindrical, conical and hyperbolic surfaces are recognizable in fig.s 1a and 1b as well in the Philips Pavilion by Le Corbusier [Capanna 2000], in the Rodin Museum in Seoul by KPF and in buildings by Gehry and Foster (figs. 1c and 1d).

Curves on domes

Two families of curves become evident from a regular parameterisation of a surface. They are called *parameter curves* and their equations are obtained by alternatively considering one of the two parameters as constant [Oprea 1997].

In the spherical surface $\mathbf{w}(t,u) = \begin{bmatrix} r\sin t\cos u \\ r\sin t\sin u \\ r\cos t \end{bmatrix}$, $(r,u) \in ([0,\pi] \times [0,2\pi])$, parallels

and meridians are curves obtained by fixing t (*u-parameter curve*) or u (*t-parameter curve*) respectively.

In the cylindrical surface $\mathbf{w}(t,u) = \begin{bmatrix} a\cos t \\ b\sin t \\ u \end{bmatrix}$, $(r,u) \in ([0,2\pi] \times R)$ and $a,b \in R^{+}$,

straight lines and ellipses are obtained with constant values of t or u respectively.

a b

Fig. 2. a) The Reichstag's Dome by Norman Foster (photo courtesy of Forster and Partners); b) The "bubble" by Renzo Piano in Turin (photo by K. Williams)

Frequently contemporary architects make evident families of *parameter curves* in vaults and domes: see for example the *Reichstag's* dome by Norman Foster in Berlin (fig. 2a) and the *bubble* of Renzo Piano in Turin (fig. 2b).

We can describe other curves belonging to a surface that are not included in the two families of parameter curves. We mention the well-known cylindrical spiral,

$$\mathbf{v}(t) = \begin{bmatrix} r\cos t \\ r\sin t \\ at \end{bmatrix}, \ t \in R, a \in R, \text{ but we can highlight other 3D-spirals developed on}$$

different surfaces.

For instance, the curve $\mathbf{v}(t) = \begin{bmatrix} rt\cos t \\ rt\sin t \\ r\sqrt{1-t^2} \end{bmatrix}$, $t \in [0,1]$, belongs to a hemi-sphere

(radius r) and its projection on the xy plane is an arc of Archimedean spiral.

In ancient monuments the spiral appears on many occasions and in different cultural contexts: examples are found in Arabian minarets and in the spire of S.Ivo alla Sapienza by Borromini in Rome.

We prefer here to focus on the use of spirals in contemporary architecture.

Norman Foster frequently uses different spirals in his projects, not only in a decorative sense but also in a functional way [Foster and Partners]: inside the dome of the Reichstag two ramps, starting from two diametrically opposite points of the floor, follow spherical spirals going up towards the top of the hemi-sphere (fig. 2a).

In the New London City Hall the conic ramps encircling the Assembly Chamber take the public from the ground floor to the viewing platform at the top of the building (fig. 3).

We'd like to mention that Foster dealt with interesting questions of optimisation in planning the City Hall. The shape is a deformation of a sphere and its surface is reduced 25% with respect to the surface of a cube having the same volume. The form was also studied to reduce the energy needs and costs as well for a better use of the sunlight [Foster and Partners].

Analysing the aspects that architects frequently deal with in designing, it becomes more and more evident that Mathematics plays a crucial role in developing an aesthetically attractive and functional construction.

Going back to the geometric aspects, cylindrical spirals appear evident in different way in modern skyscrapers: sometimes by twisting a cube (fig. 4) or the entire building, as in the project by Zaha Hadid for the trade fair area in Milan (see http://www.nuovopolofieramilano.it), other times by rotation of a motif, such as a balcony inside the Jim Mao Tower by Skidmore, Owings and Merrill in Shanghai (see http://www.emporis.com/en/wm/bu/?id=103803).

Other interesting curves can arise from the intersection of two surfaces, [Loria 1930; Loria 1925].

Fig. 3. The conic ramps of London City Hall. Photos courtesy of Foster and Partners

Fig. 4. HBS Malmö - Turning Torso Tower by Santiago Calatrava (photo courtesy of Turning Torso Meetings, http://www.turningtorsomeetings.com/)

Let us consider the curve belonging to the sphere $\mathbf{w}_1(t,u) = \begin{bmatrix} r\sin t \cos u \\ r\sin t \sin u \\ r\cos t \end{bmatrix}$,

$(r,u) \in ([0,\pi] \times [0,2\pi])$ and to the cylinder $\mathbf{w}_2(t,u) = \begin{bmatrix} u \\ \alpha + \beta \sin t \\ \beta \cos t \end{bmatrix}$,

$(r,u) \in ([0,2\pi] \times R)$, α, β being positive real parameters.

Making the choice $0 < \alpha < r/2$, $\alpha = \beta$, $\mathbf{w}_2(t,u)$ becomes a cylindrical surface passing through the centre of the sphere. The intersection of the two surfaces is the so-called *curve of Roberval*, formed by two simple closed curves (fig. 5a).

The choice $0 < \alpha < r$, $\beta = r - \alpha$ gives a cylinder tangent to the sphere. Its intersection with the sphere is called the *Hippopede of Eudoxus;* it is a closed curve having a double point where the two surfaces are tangent (fig. 5b).

The choice $\alpha = \beta = r/2$, gives a particular case (limit case of both over-mentioned). The intersection is V*iviani's curve* (fig. 5c).

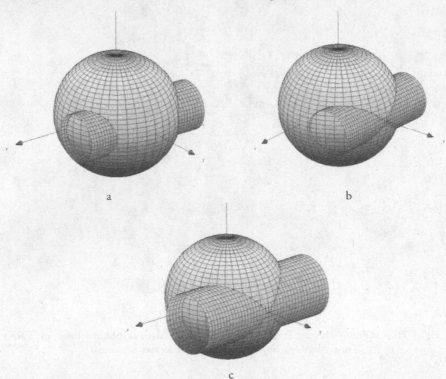

a

b

c

Fig. 5. a) Curve of Roberval; b) Hippopede of Eudoxus; c) Viviani's curve. The graphic visualizations are produced using MATLAB 6.5

These three curves are related to cutting out windows in domes, as in St. Peter's cathedral in Rome or in S. Lorenzo by Guarino Guarini in Turin.

In 1692 Vicenzo Viviani proposed the problem to construct four windows in a hemispherical vault so that the remainder of the surface can be accurately determined. Viviani's curve arises from the formation of a spherical portion, which he called *Vela Quadrabile Fiorentina* (quadrable Florentine vault), having rational measure with respect to the square of the radius: cutting out of the sphere surface the four "eyes" selected from two Viviani's curves (fig. 6a), the remaining part (two opposing *Vela Quadrabile Fiorentina*) has area equal to $8r^2$ [Marchetti and Rossi Costa 2004].

It should be noted that cutting out four equal segments of a sphere, the remaining surface is formed by two traditional sail vaults, whose area is irrational with respect to r^2 (fig. 6b).

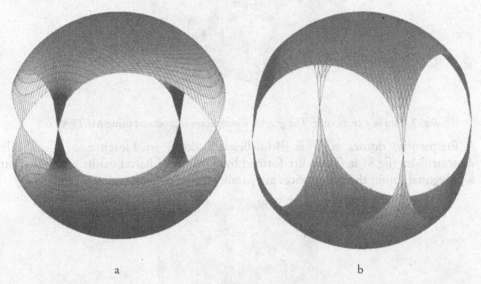

a   b

Fig. 6. The surfaces formed a) by two Vela Quadrabile Fiorentina; b) by two traditional sail vaults.
The graphic visualizations are produced using MATLAB 6.5

The intersections of two semi-cylinders (barrel vaults in Architecture), having orthogonal generatrices,

$$\mathbf{w}_1(t,u) = \begin{bmatrix} u \\ r\cos t \\ r\sin t \end{bmatrix}, (t,u) \in ([0,\pi] \times R) \text{ and } \mathbf{w}_2(t,u) = \begin{bmatrix} r\cos t \\ u \\ r\sin t \end{bmatrix}, (t,u) \in ([0,\pi] \times R),$$

are two arches of the ellipses represented by the vectors $\mathbf{v}_1(t) = \begin{bmatrix} r\cos t \\ r\cos t \\ r\sin t \end{bmatrix}, t \in [0, \pi]$

and $\mathbf{v}_2(t) = \begin{bmatrix} -r\cos t \\ r\cos t \\ r\sin t \end{bmatrix}, t \in [0, \pi]$ respectively.

These arches are called ribs or *costoloni* and they belong to cross vaults as well as to cloister vaults (fig. 7).

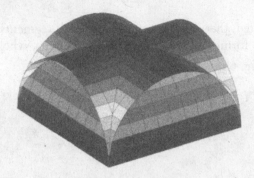

Fig. 7. Ribs in a cross vault. The graphic visualization is produced using MATLAB 6.5

Frequently, domes such as Brunelleschi's dome in Florence or the Mole Antonelliana (fig.8) in Turin are formed by portions of barrel vaults and posed on a polygonal drum; the generatrices are parallel to the polygon sides.

Fig. 8. The Mole Antonelliana, Turin (photo by K. Williams)

In Brunelleschi's dome [Battisti 1976] each cylindrical segment is composed of parts of three different cylinders with sections having separated centres, even if very close.

The vertical median section of each segment represents the directrix of the cylindrical surface; it is a polycentric curve formed by three arches of different circles, whose centres belong to the diameter of the circle inscribed in the octagonal basic section. Consequently three arches of different ellipses form the ribs of the dome of S.Maria del Fiore (fig. 9).

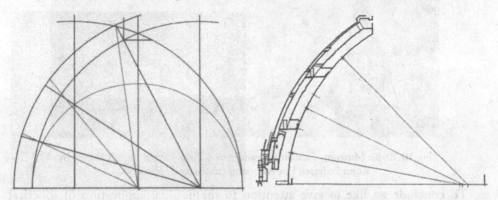

Fig. 9. Sections of the cylindrical dome segments in S.Maria del Fiore by Brunelleschi in Florence, after Battisti 1976

Among many different ancient examples of elliptical and oval domes, we would like to mention three important monuments. The Residenz Church in Würzburg by B. Neumann (1741) is one of the most important Baroque monuments in Germany [Giordano 1999]; the two churches in Rome, S. Carlino alle Quattro Fontane by Borromini and S. Andrea al Quirinale by Bernini (both seventeenth century), are also located very close to each other, so that you can enjoy the elegant architecture of both.

We cannot forget that Gaudì frequently used parabolic domes; naturally the list of citations could be longer but the choice of mentioning Gaudì, among others prominent architects, gives us the opportunity to reaffirm that his works appear to be born by improvisation although he had a strong formation in geometry.

We can collect a lot of examples of quadric surfaces in contemporary buildings.

The use of circular or elliptical cylinders is quite frequent in buildings planned by Mario Botta. Most of them are cut by an oblique plan so that the roof becomes a big round window (see http://www.botta.ch/Page/Recenti_en.php). The light entering through the glass forms a suggestive geometric pattern playing with the courses of bricks or of different stones. The effect is considerable especially in churches such as S. Giovanni Battista in Mogno and the chapel of S. Maria on Monte Tamaro (both in Ticino, Switzerland) and the Cathedral in Évry (France)

[Lavigne 2001]. The use of a conical surface is evident in the Museum Rodin by Kohn Pedersen Fox Associates (fig. 10).

Fig.10. Rodin Museum, Samsung Headquarters Plaza – Seoul (South Korea) by
Kohn Pedersen Fox Associates (photo by T. Hursley)

To conclude we like to give attention to the unusual application of spherical elements by Richard Meier in the church of Tor Tre Teste, Rome (fig. 11). Three portions of a sphere, having equal radius and centred on the same line, form the main part of the church *Dives in Misericordia*. These geometric portions, called *sails* by Meier himself, are walls and domes at the same time [Meier 1997]. They are self-supporting and give the building an impressive symbolic as well as religious character.

Fig. 11. The Church Tor Tre Teste, Rome (photo A. Geroldi)

The concrete panels (*conci*) forming the structure are studied with regard to modern environmental impact, so that they have particular qualities. They are not only more resistant but are also designed to maintain their white brilliant colour through the addition of Carrara marble powders and titanium bi-oxide.

Conclusions

We appreciate this new trend in building, not only because of the sociological, ecological, energetic aspects, but also because this kind of planning makes evident the strong contribution of mathematical modelling. Thus the creations of contemporary architecture are strictly bound up with geometry and with the more sophisticated applications of the techniques of modern mathematics.

Acknowledgment

The authors wish to thank the following for generous permission to reproduce images: Foster and Partners (figs. 2a, 3a, 3b); Kohn Pederson Fox Associates (fig. 10); HBS Turning Torso Meetings (fig. 4).

Notes

1. Richard Meier, the famous architect born in the USA (New Jersey) in 1934, commenting on his project for the third Millennium Church, built in Tor Tre Teste (Rome, 1998-2003).

2. $C^1(I)$ denotes the vector space of continuous functions with derivative function continuous in the interval I; the components of the vector v(t) belong to this functional space. The interval I may be closed, open or unlimited.

3. The symbol × means Cartesian product between intervals. The vector product is indicated by the symbol ∧.

References

BATTISTI, E. 1976. *Filippo Brunelleschi*. Milan: Electa Editrice.

CAPANNA, A. 2000. *Le Corbusier Padiglione Philips.* Turin: Universale di Architettura.

FOSTER AND PARTNERS. http://www.fosterandpartners.com.

GIORDANO, A. 1999. *Cupole, volte e altre superfici.* Turin: Utet.

HÖWELER, E. 2004. *Skyscraper: Vertical Now.* New York: Universe Publishing.

LAVIGNE, E.2000. *EVRY La cathédrale de la Résurrection.* Paris : Edition du Patrimoine.

LORIA, G. *Curve sghembe speciali,* Bologna: N. Zanichelli Ed..

———. 1930. *Curve piane speciali.* Milan: U. Hoepli Ed.

MARCHETTI, Elena and Luisa ROSSI COSTA. 2004. Mathematical and historical investigations on domes and vaults. Pp. 73-80 in the Proceedings of the Dresden International Symposium of Architecture, R. Weber and M. Amann eds. Dresden.

MARIO BOTTA ARCHITETTO. http://www.botta.ch.

MEIER, Richard. The Church of the Year 2000. Pp. 13-23 in *Proceedings of Workshop "Materie e Strutture per il Nuovo Millennio"*, Milan Polytecnic - 25 February 1997, Milan.

OPREA, J. 1997. *Differential Geometry and its applications.* New York: Prentice Hall.

About the authors

Elena Marchetti received her doctorate in Mathematics at the Faculty of Sciences at the Università degli Studi in Milan. She was a researcher of Mathematical Analysis at the Department of Mathematics of the Politecnico of Milan, and since 1988 is an associate professor of Istituzioni di Matematica at the Faculty of Architecture of the Politecnico of Milan. For many years she taught in courses of Mathematical Analysis to engineering students, and since 1988 she has taught Mathematics courses to architecture students. Her research activity is concentrated in the area of Numerical Analysis, principally regarding numerical integration and its applications. She has produced numerous publications in Italian and international scientific journals. Her participation and collaboration in several conferences dedicated to the application of mathematics to architecture has stimulated her interest in this subject. The experience gained through intense years of teaching courses to architecture students has led her to publish some textbooks, one of which regards lines and surfaces and has a multimedia support package, on the production of which she collaborated. She published papers with the aim to connect arts, architecture and mathematics.

Luisa Rossi Costa earned her doctorate in Mathematics in 1970 at Milan University and she attended lectures and courses at Scuola Normale Superiore in Pisa and at Istituto di Alta Matematica in Rome. Since October 1970 she has taught at the Engineering Faculty of the Politecnico of Milan, where she is associate professor of Mathematical Analysis. She first developed her research in Numerical Analysis, on variational problems and on calculating complex eigenvalues. Her interest then changed to Functional Analysis and to solving problems connected with partial differential equations of a parabolic type. She also studied inverse problems in order to determinate an unknown surface, an unknown coefficient in the heat equation and a metric in geophysics, with the purpose to find stable solutions in a suitable functional space. She published several papers on these subjects. She took part in the creation of lessons for a first-level degree in Engineering via the Internet. She also researches subjects regarding teaching methods and the formation of high school students; she collaborates on the e-learning platform M@thonline. Following a continuing interest in art and architecture, and believing that mathematics contains a strong component of beauty, she tries to connect these apparently different fields. She published papers connected to this aim.

J. Iñiguez
A. Hansen
I. Pérez
C. Langham
J. Rivera
J. Sánchez
J. Acuña

Cochise College,
Douglas Campus
4190 W Highway 80
Douglas AZ 85607 USA
iniguezj@cochise.edu,
iniguez@c2i2.com

keywords: algebra, division
in extreme and mean ratio,
golden mean, golden
rectangle, golden trapezoid

Research

On Division in Extreme and Mean Ratio and its Connection to a Particular Re-Expression of the Golden Quadratic Equation $x^2 - x - 1 = 0$

Abstract. The golden quadratic $x^2 - x - 1 = 0$, when re-expressed as $(x)(1) = 1/(x-1), x = 1.618$, can be interpreted as the algebraic expression of *division in extreme and mean ratio* (DEMR) of a line of length $x = 1.618$ into a longer section of length 1 and a smaller of length $(x-1)$. It can, however, also be interpreted as the formulation of the area of a golden rectangle of sides $x = 1.618$ and 1, and as the system of equations constituted by $y = x$, and $y = 1/(x-1)$. Based on the well-known connection existing between the first two of these interpretations, the authors address the problem of finding out the thread connecting the golden rectangle with the system of equations referred to above. The results obtained indicate first that this system, like the golden rectangle, also carries in its geometry the essential traits of DEMR; and, second, that it implicitly subsumes the simpler rectangular geometry of its alternative interpretation. The process of developing these connections brought forward a heretofore apparently unreported golden trapezoid of sides $\Phi, 1, \phi$, and $\sqrt{2}$.

I. Introduction

This number is no other but the ratio known as the aurea sectio, which has played such a role in attempts to reduce beauty of proportion to a mathematical formula.

H. Weyl

The golden ratio. It is in Euclid's *Elements* where the first appearance/definition of the problem of "division in extreme and mean ratio" (abbreviated as DEMR), later to be known as the "golden section", "golden ratio", or "divine proportion", can be found [Herz-Fischler 1998, vii; Livio 2003, 3]. There, this problem is discussed both from an area perspective: "To cut a given straight line so that the area of the rectangle contained by the whole line and one of the segments is equal to the area of the square on the remaining segment" [Euclid 1956, I, II, prop. 11, 402; Herz-Fischler 1998, 1], as well as from a line sectioning approach [Euclid 1956, 2, VI, prop. 30, 267], a problem defined by Euclid as follows: A straight line is said to have been cut in extreme and mean ratio when, as the whole line is to the greater segment, so is the greater to the less" [Euclid 1956, 2, VI, def. 3, 188]. For a line of length x divided into a long section of length 1 and a short section of length

$(x-1)$, this condition can be expressed as $x/1 = 1/(x-1)$ or, through rearrangement, as $x^2 - x - 1 = 0$. When this equation is solved, we get two irrational numbers as roots: $x = (1+\sqrt{5})/2$, and $x = (1-\sqrt{5})/2$, corresponding, respectively, to the rounded numbers $x = 1.618$, and $x = -0.618$. The fact that length is the magnitude being considered here conveys physical significance only to the positive root. If so, to divide then a line of length 1.618 in extreme and mean ratio, two sections need to be produced out of it, one of length 1 and the other of length 0.618, numbers that bring compliance with the previously stated condition in the following form: $1.618/1 = 1/0.618 = 1.618$. We can now take any of those two segments and repeat with it the previous procedure. If the segment chosen for further division is that of length 1, we can then set up equation $1/x = x/(1-x)$, with x now being the largest section into which a line of length 1 has to be divided to comply with the golden ratio. Solving it produces the following result: $x = (-1+\sqrt{5})/2 = 0.618$. If so, the shortest section will be of magnitude 0.382, and again, the golden ratio materializes in the number 1.618, as follows: $1/0.618 = 0.618/0.382 = 1.618$. At least in principle, this procedure can be repeated ad infinitum, and every time, the mentioned ratios will invariable produce the number 1.618, or as expressed in a medieval edition of Euclid..."Whatever happens to one line divided according to EMR is proved to happen to every line likewise divided" [Herz-Fischler 1998, xx]. It is on reason of it quantifying the process of division in extreme and mean ratio that the number 1.618 is called the golden mean, or golden ratio. Even if several denominations are in use in the literature to designate it, here it will be represented with a capital Phi (Φ), while for its inverse, 0.618, the symbol to be used is lower case phi (ϕ).

The golden rectangle. According to Greenberg: "The number... $(1+\sqrt{5})/2$ was called the *golden ratio* by the Greeks, and a rectangle whose sides are in this ratio is called a *golden rectangle*" [Greenberg 1980, 28]. Zippin, on the other hand, defines the shape of the golden rectangle in terms of the previously stated golden ratio preserving property of division in EMR: "*If one cuts a square away from it then the rectangle that remains has exactly the same shape as the original rectangle. By 'same shape' we mean that the ratio of shorter side to longer side is the same, i.e., that the two rectangles are similar*" [Zippin 2000, 75]. The existence of this procedure allowing for the replication of the original rectangle into similar rectangles, has led Zippin to dub this quadrilateral the "self perpetuating golden rectangle" [Zippin 2000, 76]. These two traits associated to every geometrical figure quoted as 'golden' in the relevant literature – a) contiguous sides conforming to the golden ratio; and b) The existence of a self replicating mechanism – appears to be the standards against which any new claim of "goldenness" has to be measured.

Golden polygons. The irregular polygons commonly referred to as "golden" in the refereed literature and/or books on the subject, are the following:

– The two isosceles triangles subsumed by the regular pentagon [Huntley 1970, 24; Kappraff 1991, 87; Sharp 2002].

– A right triangle referred to by Sharp as *Escher's Golden Section triangle* [Sharp 2002].

Let us note in this regard that the sole appearance of Φ or ϕ in a given figure does not automatically convey the 'golden' qualifier, as above understood. Consider, for example, the rectangle with sides Φ and ϕ. Notwithstanding the fact that these are golden numbers, this rectangle cannot be considered golden because its sides do not conform to the golden ratio.

In addition to irregular polygons, a golden conic section [Huntley 1970, 65] and a golden polyhedron are found in the literature [Huntley 1970, 96]. A discussion of these falls, however, beyond the scope of this paper.

II. Argument

Antecedents. In the course of a parallel research project involving symmetry, our student research group became aware of the fact that different geometries may arise when a given equation is re-expressed in a number of algebraically equivalent forms. For example, the golden quadratic: $x^2 - x - 1 = 0$, as written, can be interpreted as a system composed of $y = x^2 - x - 1$ and $y = 0$. When expressed in any of the following forms, however, $a)\ x(x-1) = 1$; $b)\ (x)(1) = 1 \setminus (x-1)$; $c)\ (x-1)(1) = 1/x$; $d)\ x^2 - x = 1$; $e)\ x^2 = x + 1$; the interpretation changes, respectively, to (a) the formulation of the area of a rectangle of sides x, and $(x-1)$; (b) either the formulation of the area of a rectangle of sides x and 1, or the system composed of $y = x$, and $y = 1/(x-1)$; (c) either the formulation of the area of a rectangle of sides $(x-1)$ and 1, or the system consisting of equations $y = x - 1$ and $y = 1/x$; (d) the system composed of $y = x^2 - x$ and $y = 1$, and finally, (e) the system composed of $y = x^2$ and $y = x + 1$. This realization led our student research group to inquire about the connection existing, if any, between those different geometries. The first step in this endeavor, the subject matter of this paper, deals with identifying the connection existing between the alternative interpretations contained within case (b) described above.

A hidden connection made visible. Two questions propel this work: In which way does DEMR, whose algebraic expression is the golden quadratic, permeate the geometries of the different re-expressions of such an equation? and What connection can possibly exist between a golden rectangle of sides $x = 1.618$ and 1, and the graph of an equilateral hyperbola unfolding around the vertical asymptote $x = 1$, and intersected by the straight line $y = x$? In regard to this last question, the fact that the simpler of these two geometries is that of the golden rectangle led us to re-formulate the second question in the following terms: Does the graphical solution of the system $y = x$, $y = 1/(x-1)$ subsume in any way the said golden rectangle of sides $x = 1.618$ and 1, and area $(x)(1) = (1.618)(1) = 1.618$?

The first step taken in our endeavor to answer these questions was to make the geometry of this system explicit by graphing it. This graph is depicted in fig. 1.

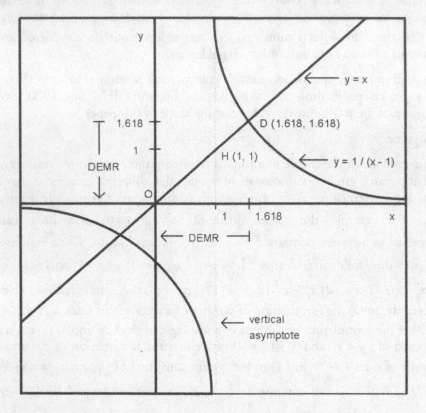

Fig. 1. Graphical solution for the system $y = x$ and $y = 1/(x-1)$. The projection of diagonal OD on the coordinate axes materializes on them the DEMR process for a line of length 1.618.

Once done, and for the reason previously advanced regarding the requirement of positive numbers for the lengths of the sides of the golden rectangle, we decided to concentrate our attention on the first quadrant portion of this graph. This proved to be a good decision because in looking at it, we became immediately aware of the fact that the two numbers quantifying the sides and consequently the area of this golden rectangle – 1 and 1.618 – appeared as the coordinates of the intersections of the line $y = x$ with, first, the vertical asymptote $x = 1$, and second, with the first quadrant portion of the graph of $y = 1/(x-1)$. These two intersections, labeled H and D, define, in turn, two 45° diagonals: OH and OD. In pondering the significance of these diagonals, we eventually realized that when projected on the coordinate axes they reproduced DEMR in its entirety with a line of length 1.618 and its golden sections of lengths 1 and 0.618. At difference of the well known fact that the length of the sides of the golden rectangle are in golden ratio, here we found DEMR as a whole.

Having found the answer to the first of the two questions propelling this work, we continued pondering about the stated diagonals. The fact that those were 45° diagonals, combined with the fact that any such diagonal implies a square, prompted us to draw the squares OAHG and OBDF and, by extension, square HCDE, as depicted in fig. 2.

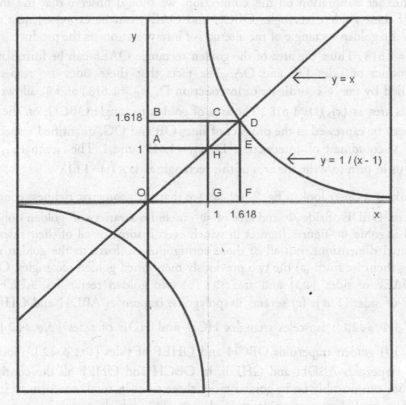

Fig. 2. Squares OBDF, OAHG and HCDE, implicitly present in fig. 1 through diagonals OD, OH and HD, have been explicitly depicted here. Their superimposition on fig. 1 brings forward a polygonal richness not apparent in their absence. Among the figures bound by square OBDF we find golden rectangles OBCG and OAEF, identical to that associated with the alternative interpretation of this system, as well as golden trapezoids OBCH and OHEF

Once those squares were drawn, an unexpected geometrical richness came to life. It was in this polygonal richness bound by square OBDF that we found, in the form of polygons OBCG and OAEF, the golden rectangle of sides 1.618 and 1, associated to the alternative interpretation of this particular re-expression, confirming our original suspicion that the complex geometry of $y = x$, $y = 1/(x-1)$ does implicitly subsume the simpler alternative interpretation.

In a Platonic sense, intersections H and D, of respective coordinates (1, 1) and (Φ, Φ), as well as diagonals OH and OD defined by them and constituting the origin of the argument unfolding here, can be considered as shadows on the surface of the graph hinting at a reality richer in forms and complexity. This reality becomes apparent when the source of those shadows, squares OAHG and OBDF, are drawn on the graph.

As further elaboration on this connection, we should observe that the areas of both of these golden rectangles OBCG and OAEF, can be expressed, just as the area of the golden rectangle of the alternative interpretation, as the product: $(x)(1)$, with $x = 1.618$. Thus, the area of the golden rectangle OAEF can be formulated as the product of sides OF and OA. The facts that these sides are respectively quantified by the x coordinate of intersection D, $x_D = 1.618$, and 1, allows us to write its area as $(x_D)(1) = 1.618$. The area of golden rectangle OBCG, on the other hand, can be expressed as the product of sides OB and OG, quantified respectively by the y coordinate of intersection D, $y_D = 1.618$, and 1. The fact that $y_D = x_D$ allows us in turn to write the area of this rectangle as $(x_D)(1) = 1.618$

In taking a closer look at fig. 2, we can see that the geometric richness contained by square OBDF of side Φ and area of Φ^2 includes a variety of golden polygons, as well as some Φ-figures (figures in which even if some or all of their sides have Φ-related dimensions, not all of those contiguous conform to the golden ratio). Among them we find: (a) the two previously mentioned golden rectangles, OBCG and OAEF of sides $[\Phi, 1]$ and area Φ; (b) two golden rectangles, ABCH and GHEF of sides $[1, \phi]$; (c) several Φ-polygons: trapezoids ABDH and GHDF of sides $[\phi, \Phi, \phi\sqrt{2}, 1]$; isosceles triangles HCD and HDE of sides $[\phi, \phi, \phi\sqrt{2}]$ and, finally, (d) golden trapezoids OBCH and OHEF of sides $[\Phi, 1, \phi, \sqrt{2}]$. Note that unlike trapezoids ABDH and GHDF, in OBCH and OHEF all the contiguous sides that can possibly be in golden ratio, three sides in total, conform to it. The area of each of these trapezoids is equal to $\Phi - 1/2$, which can also be written as $\phi + 1/2$ or as $(\Phi + \phi)/2$. This last expression allows us to realize that each of those trapezoids is bounded by two golden rectangles of consecutive dimensions: OBCG and ABCH, and as such can be considered as the golden polygon arising in the intermediate step of transiting from one to the other when, instead of removing the square OAHG from OBCG to produce ABCH, we remove as a first step the isosceles right triangle OHG. This realization leads us to the replication mechanism of the golden trapezoid shown in Figure 3. With the golden trapezoid OBCH as a starting point, we proceed to remove from it triangles OAH and JCH in order to generate trapezoid ABJH. From it we in turn remove triangles KJH and LBJ to produce trapezoid ALJK, and so on ad infinitum. The zigzag spiral characteristic of these replication mechanisms is apparent in the figure.

Let us now make clear that when it is stated that those golden trapezoids have been up to date unreported as such, it is meant unreported as *golden* figures. Certainly, those trapezoids can be found in Euclid [1956, III, xiii, 447] and Zippin [2000], among others. No reference qualifying these as 'golden' has been found in the relevant literature.

It should be recognized that in figures such as a right triangle or the golden trapezoids discussed above, the number of contiguous sides complying with the golden ratio is subject to fundamental restrictions originating in their own geometry. For example, if a, b, and c are the sides of a right triangle with a and b conforming to the golden ratio, it can then be shown through a very simple exercise in elementary trigonometry that due to the connection established between them by Pythagoras's theorem, no two other sides can conform to it. These same considerations apply to the golden trapezoids here being introduced. This is the reason behind the fact that in these golden trapezoids only three contiguous sides conform to the divine proportion.

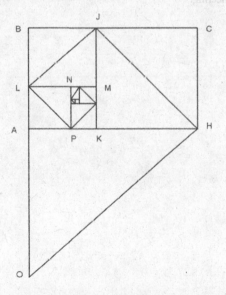

Fig. 3. The removal of isosceles right triangles OAH and JCH from trapezoid OBCH produces similar trapezoid ABJH. If from it triangles KJH and LBJ are now removed, similar trapezoid ALJK appears and so on, ad infinitum. In any of these golden trapezoids, the contiguous orthogonal sides are in golden ratio

III. Conclusion

The results herein obtained indicate that the golden quadratic embeds the geometries of its different representations with the traits of division in EMR. When the re-expression $(x)(1) = 1/(x-1)$ is seen as the system of equations $y = x$, $y = 1/(x-1)$, it was found that its geometry not only projects on the coordinate axes the process of DEMR in it entirety, it also implicitly subsumes the simpler rectangular geometry of its alternative interpretation.

References

EUCLID. 1956. *The Elements*. Thomas L. Heath, trans. 2nd ed. New York: Dover.

GREENBERG, M. 1980. *Euclidean and Non-Euclidean Geometries*. San Francisco: W H Freeman.

HERZ-FISCHLER, R. 1998. *A Mathematical History of the Golden Number*. New York: Dover.

HUNTLEY, H. 1970. *The Divine Proportion*. New York: Dover.

KAPPRAFF, J. 1991. *Connections: The Geometric Bridge between Art and Science*. New York: McGraw-Hill.

LIVIO, M. 2003. *The Golden Ratio*. New York: Broadway Books.

SHARP, John. 2002. Spirals and the Golden Section. *Nexus Network Journal* 4, 1: 59-82. http://www.nexusjournal.com/Sharp_v4n1-intro.html

ZIPPIN, L. 2000. *Uses of Infinity*. New York: Dover.

About the authors

José C. Iñiguez (M.Sc. physical chemistry, Cinvestav, 1973) works as a science and math instructor at Cochise College. His coauthors in this paper, members of Cochise's student research group, are all students of science or engineering.

L. Consiglieri

Mathematics Dep/FCUL
University of Lisbon
1749-016 Lisbon, Portugal
lcconsiglieri@fc.ul.pt

V. Consiglieri

Faculty of Architecture
Technical University of Lisbon
Lisbon, Portugal

keywords: phenomenological
forms, rhythm, continuous
functions

Research

Structure of Phenomenological Forms: Morphologic Rhythm

Abstract. The images in architecture are handed down through mathematical forms. The meaning of the plastic value of the forms and the conflict between their visual boundaries are a result of the geometrical composition of the object. Since Stonehenge in Britain, the Egyptian pyramids, the Greek Parthenon or the Roman Pantheon, architecture has been a reflex of simple boundaries without accidental confrontations. Nowadays materials are organised through movement/change in order to represent the required profile. This developed structure emerges in the artistic manifestations according to the theory of continuity. As an expression of the formal quality in opposition to the ancient characteristics of quantity, a new conception of rhythm appears. The concept of a cell as an architectural element that can have any biological form and can be grouped itself according to different ways and functions (such as repetition of floors) is introduced. This concept of cell permits eurhythmy (harmony in the proportion of a building) through the notion of rhythm once all the elements of a building are situated among themselves.

1 Introduction

Throughout the last fifty years of the twentieth century there was been an intense engagement closely connected to the research of artistic composition, in particular architecture, by integrating universal scientific knowledge in fields such as mathematics, physics, and biology into the creative work [Kepes 1968]. Indeed, a change of attitude in Modern Art appeared: to serve society we could no longer understand the laws of modern art without an awareness of notions about space/time [Giedion 1969]. The re-examination of cultural modernity and the influence of architectural linguistics such as semiotics and structuralism increased. Also, sensibilities changed and new cultural forms attempted an approach to other typologies of emotion [Consiglieri 2000]. That transformation of knowledge integrates the contemporary science of composition. Communication theories and phenomenology present additional methods, exhibit alternative solutions based on observation, and offer new thought paradigms for the crisis of meaning within architecture. What structures allow the understanding of a form of expression? Can there be meaning in form or only in content? Can it be sustained through innovation and invention?

In *Gestalttheorie* [Guillaume 1937], an object is developed according to the inherent character of the various surfaces placed in a given relationship to the volume, namely the regular faces, vertices, edges, angles and orientations necessary for the basic views and plans. The phenomenological forms are structures made of

---

1590-5896/06/020101-10 DOI 10.1007/s00004-006-0022-x

sets of cells grouped together by a process of addition; these arrangements would serve different shapes of rooms or plans of the building [Steadman 1983]. The application of this methodology in the construction of a building corresponds to the production of modular elements and their correlations. The relationship of these small pieces or cells depicted on rhythmic grids is the fundamental element that defines the whole [Bonomi 1974].

In the present work, we consider cells for habitation that are modular structures oriented so as to enclose different spaces and constitute one unity of the space organised in such way that the position of the cell can be recognised, and they are structured in a continuous movement which produces a rhythm.

For a long time rhythm was thought of as a cadence of measures between the different parts and between these and the whole. Rhythm was in relation to the proportion of the human body, which was considered by the Greeks as *analogy*. Hence each part of the building was in fitting proportion to the height, width and depth, and all parts were in their own place within the symmetric whole of the building. This sequence of proportions or figurative relations created the notion of rhythm. The periodic character introduced according to scale defined the rhythm of the composition. Moreover, rhythm was connected with the notions of numbers by their capacity to create an Order [Ghyka 1978]. Thus rhythm was defined as a conceived periodicity [Servien 1930]. Thus, rhythms were considered by means of the regularity relation between space and time.

Through periodic and non-periodic modular concepts introduced by Roger Penrose, the object became based on mechanical strategies, taking into account that the rhythm of an architectural composition is still equilibrium, the element which organizes the relationship between their parts in a harmony or in a dissonance [Darvas 2003]. In recent years, there appeared a new interpretation of the world: the *emergence* [Weinstock 2004]. It is in this state of art that structural morphologies concerning *growing* architecture deal with the organising, shaping, building and transforming new configurations from the manipulation of unit elements, via algorithms [Leoz 1969] (for instance, the Rose and Stafford algorithm [Alsina and Trillas 1984, 154;162]), genetic codes [Lalvani 1982] or group actions [Schattschneider 2004].

The principal aim of the current work is to express the architectural processes through continuum rules. In the next section, the concept of morphologic rhythm as an architectural cell is defined, and some mathematical examples are presented, demonstrating the advantages of open-source computer programs. Section 3 is devoted to a discussion of some architectural examples.

2 Morphologic rhythm

The classical concept of rhythm, which consists in the ordered repetition of isometries (i.e., translations, rotations and reflections), can be generalized to a

---

morphologic one that emerges as a self-organized life cell. Shape changes also can occur.

Let C be a *cell* defined by

$$C := \{(x, y, z) \in IR^3 : f(x, y, z; t) \le 1\},$$

where x, y, z represent (as usual) the spatial Cartesian coordinates and f denotes a function defined in the three-dimensional Euclidean space and depending on a time parameter $t \in [0, +\infty[$. For instance, if we take $f(x, y, z; t) = x^2 + y^2 + z^2$, the set C reads

$$C := \{(x, y, z) \in IR^3 : f(x, y, z; t) = x^2 + y^2 + z^2 \le 1\}.$$

Thus it can be called a *steady state spherical cell* since f is constant as a function of the parameter time t and the set C represents a closed ball of radius 1 that is the union of the boundary (sphere) and its interior (fig. 1).

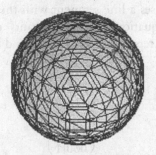

Fig. 1. Graphical representation of a steady state spherical cell of radius 1. Graphic visualization produced using Maple 6

We say that a cell C has a *morphologic rhythm* if

$$\exists R > 0, \quad C(t) \subseteq B_R(0), \qquad \forall t \in [0, T],$$

where $B_R(0)$ denotes an open ball centered at the origin (0,0,0) with radius $R > 0$, f is a continuous function defined on $IR^3 \times [0, T]$ and $T > 0$. The meaning of the above definition can be understood as that the rhythm will take part of the lifetime of the cell. The requirement that the cell belong to an open ball for every time parameter provides the boundedness of the final pattern, since the pattern solutions that blow up have no physical (architectural) meaning.

Let us first consider the kinematics of the center of mass of a cell in order to recognise the classical concept of rhythm, that is, its rigid motion through discrete different instants of time. For the spherical cell of radius 1 (fig. 2) given by

$$C := \{(x, y, z) \in IR^3 : f(x, y, z; t) = x^2 + (y - t)^2 + z^2 \le 1\} \qquad (1)$$

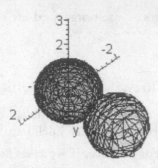

Fig. 2. Graphical representation of a spherical cell of radius 1 at instants of time t = 0 and t = 2 when $f(x, y, z; t) = x^2 + (y - t)^2 + z^2$. Graphic visualization produced using Maple 6

the classical notion of rhythm corresponds to consider instants of time whose the difference between them is greater or equal than twice the radius of the cell, that means, the cell behaves like a rigid ball which moves by translation. Here the cell appears as invariant under translation. Indeed a rigid and uniform motion in which each point of the cell traverses a line segment with the same length and direction. For instance, taking in equation (1) the instants of time t=0 and t=2, the translation has the algebraic representation [Alsina and Trillas 1984, 142]

$$\left(\begin{array}{ccc|c} 1 & 0 & 0 & 0 \\ 0 & 1 & 0 & 2 \\ 0 & 0 & 1 & 0 \\ \hline 0 & 0 & 0 & 1 \end{array} \right).$$

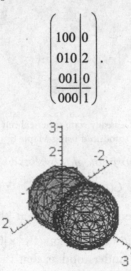

Fig. 3. Graphical representation of a spherical cell of radius 1 at instants of time t = 0 and t = 1 when $f(x, y, z; t) = x^2 + (y - t)^2 + z^2$. Graphic visualization produced using Maple 6

However, if different instants of time can be taken into account, a translation in the broad sense can be stated. Indeed, fig. 3 illustrates the spherical cell of radius 1 defined as in (1), that is, it can be regard as moving through the same trajectory $u : IR \rightarrow IR^3$ defined by

$$u(t) = t(0, 1, 0) = (0, t, 0),$$

but now at instants of time t=0 and t=1. Then there exists an intersection of the cell with itself. Fig. 4 illustrates the spherical cell of radius 1 moving through the trajectory $u : IR \rightarrow IR^3$ defined by $u(t) = t(-1, 1, 1) = (-t, t, t)$.

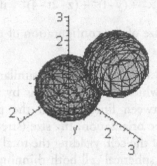

Fig. 4. Graphical representation of a spherical cell of radius 1 at instants of time t = 0 and t = 1 when $f(x, y, z; t) = (x + t)^2 + (y - t)^2 + (z - t)^2$. Graphic visualization produced using Maple 6

Other modular coordination created with isometries as generators can be performed for generating decorative patterns as well as architectural phenomenological forms.

If the center of the spherical cell at instant time t = 0 is at the origin and its movement describes the parabola $p(t) = -2t^2 + 4t$, in the plane 0yz, then the visual interpretation of the morphologic rhythm corresponds to the movement through a nonlinear trajectory (fig. 5, at three distinct instants of time).

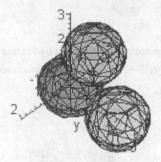

Fig. 5. Graphical representation of a spherical cell of radius 1 at instants of time t =0, t = 1 and t = 2 when $f(x, y, z; t) = x^2 + (y - t)^2 + (z + 2t^2 - 4t)^2$ for $0 \le t \le 2$.

This movement cannot be performed even more by a simple composition of translations and rotations. Only when using discrete instants of time, nonlinear trajectories can be reduced by plotting to linear functions at the chosen discrete instants of time. For instance, the above parabola can be interpolated by two different linear functions $u:[0,1] \to \mathrm{IR}^3$, $u(t)=t(0,1,2)$, and $v:[1,2] \to \mathrm{IR}^3$, $v(t)=(0, 1, 2) + (t-1)(0, 1, -2) = (0, t, -2t+4)$ and the cell is then defined by the function

$$f(x,y,z;t) = \begin{cases} x^2 + (y-t)^2 + (z-2t)^2 & \text{if} \quad 0 \le t \le 1 \\ x^2 + (y-t)^2 + (z-2t-4)^2 & \text{if} \quad 1 < t \le 2. \end{cases}$$

Here the shape and the size of the configuration of the cell are not altered, only its position is changed.

This new concept of rhythm also includes the similarity notion, i.e., the result of contractions or dilatations which alters all lengths by a same factor but preserves the measures of angles between lines. Indeed the present concept allows the alteration of the location, the orientation, the size (stretching or thinning) and the shape of the configuration of the cell, yielding the total change of the initial format of the cell. Fig. 6 shows the spherical cell both thinning and shape changing at the instant of time t = 1. Finally, the cell motility involving translocation of the cell centroid as well as changes or distortions in the cell shape is shown at fig. 7. Then the outcome of the morphologic rhythm is the continuum process set up by the cell as a function of the parameter time and its connection with the biological forms.

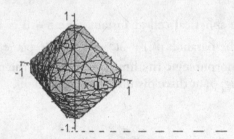

Fig. 6. Graphical representation of a spherical cell of radius 1 at instant of time t = 1 when $f(x, y, z; t) = |x|^{2-t} + |y|^{2-t} + |z|^{2-t}$. Graphic visualization produced using Maple 6

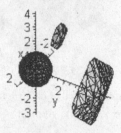

Fig. 7. Graphical representation of a cell of radius 1 at instants of time t =0, t = 1 and t = 2 when

$$f(x, y, z; t) = \left(|x|^{t^2/2-3t/2+2} + |(2t+1)y - 5t^2|^{t^2/2-3t/2+2} + |z - 3t(2-t)|^{t^2/2-3t/2+2} \right) / (1 + t(t-1))$$

3 Discussion and conclusions

Taking a cell as the basic object from which are developed different variations, we can coordinate the compositional process so as to obtain a determined grid that allows the effective rhythmic continuation and a flexible figurative relationship between the different cellular components (fig. 8). Thus the cell is the architectural measure denoting both discipline and conceptual freedom.

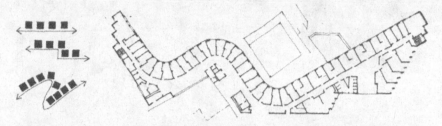

Fig. 8: Morphologic rhythm/Alvar Aalto, MIT, Cambridge USA

Upon these co-ordinator modulations, where an element assumes different directions in plan or generally in 3D space, the movements of growth are articulated such that they can be conditioned to have no boundary (figs. 9 and 10). This growth in composition that is not subjected to any modular limit is a solution for an open, free space building that expands through axial elements.

Contrary to this open scheme, Frank Lloyd Wright studied grids of rotative symmetrical cells in order to obtain a dense building (fig. 11). Since the process of grids to form composition in space by Wright rose from a constant geometrical modulation on which the form develops, it also can be assumed to be formed from morphologic rhythm. If those habitational cells of one floor were to have a standard plan shape, the habitability is restricted to the situations of fulfilment of the interiors as well as the systems of lighting and ventilation. If those habitational cells were to have a plan shape which is made by a continuous rhythmic system, the fulfilment of habitat housing would be totally different of the former, affording more interior spaces.

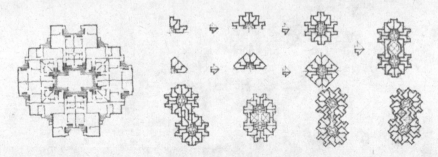

Fig. 9: Boffil, study for Gaudi Quarter

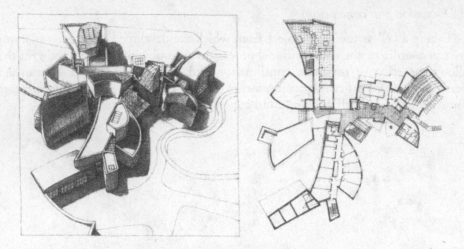

Fig. 10: Frank Ghery, EMR, Germany

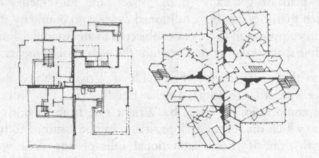

Fig. 11: F.L. Wright: Suntop House, Pennsylvania / St. Marks Tower, NY

Fig. 12: Arata Isozaki, study for a spatial city

To comprehend the great complexity of the rhythm under one only element, consider the study of Arata Isozaki for a spatial city: repeatedly fitting the given element results in a (non-unique) solution in the set of the potential solutions of different symmetric and asymmetric shapes (fig. 12).

Despite the cells being conditioning elements on the organization of the spaces, the house must not be thought as a machine of habitat, but constitute a private place inserted in an urban space, where the family is joined in a community. This may be called the *poetics of habitat*, where the closeness allows intimate relationships within the basic laws of proximity and centralization. Moreover, like biological cells (Lat. *cella* "concealed" storage or living room) the architectural cells can perform several kinds of motion that continuously shape their structure in relation to and in interaction with a varying environment. It is the dichotomy of cells, existing and moving either as autonomous units or as components within a multicellular tissue, that leads to a final question: what are the conditions and principles of tissue formation for eurhythmy?

Acknowledgment

The authors are deeply grateful to Teotónio Agostinho for figs. 8-12.

References

ALSINA, C. and E. TRILLAS. 1984. *Lecciones de Algebra y Geometría, Curso para estudiantes de Arquitectura*. Barcelona: Ed. Gustavo Gili, S.A.

BONOMI, A. 1974. *Fenomenologia e estruturalismo*. São Paulo: Editora Perspectiva.

CONSIGLIERI, V. 2000. *As significações da arquitectura 1920-1990*. Lisbon: Ed. Estampa.

DARVAS, G. 2003. Perspective as a symmetry transformation. *Nexus Network Journal* 5, 1: 9-21.

GHYKA, M.C. 1978. *El número de oro. I los ritmos, II los ritos*. Barcelona: Ed. Poseidon.

GIEDION, S. 1969. *Espacio, Tiempo y Arquitectura*. Barcelona: Dossat.

GUILLAUME, P. 1937. *La psychologie de la forme*. Paris: Flammarion, éditeur.

KEPES, G. (ed). 1968. *Module, Proportion, Symétrie, Rythme*. Bruxelles : La Connaissance. (English version, New York: George Braziller, 1966.)

LALVANI, H. 1982. *Structures on hyper-structures: multi-dimensional periodic arrangements of transforming space structures*. New York: H. Lalvani.

LEOZ, R. 1969. *Redes y ritmos espaciales*. Madrid: Ed. Blume.

SERVIEN, Pius. 1930. *Les rythmes comme introduction physique à l'Esthetique*. Paris: Bibl. de la Rev. des Cours et Conférences,

SCHATTSCHNEIDER, D. 2004. *M.C. Escher: visions of symmetry*. 2nd ed. New York: H.N. Abrams.

STEADMAN, J.P. 1983. *Architectural morphology. An introduction to the geometry of building plans*. Pion Limited, London.

WEINSTOCK, M. 2004. Morphogenesis and the mathematics of emergence. In "Emergence: morphogenetic design strategies", *Architectural Design* 74, 3 (May-June 2004): 10-17.

About the authors

Luisa Consiglieri received her B.S., M.S. and Ph.D. degrees in mathematics from University of Lisbon in 1988, 1992 and 2000, respectively. She has been at the Faculty of Sciences of University of Lisbon since 1987 and is currently Assistant Professor. Her research interests span a variety of areas in differential equations such as existence, uniqueness and regularity of solutions, fluid mechanics, electromagnetism, biomechanics, and applications to biomedical problems. At the present her principal aim is interdisciplinarianism.

Victor Consiglieri received a B.S. degree in architecture from Escola Superior das Belas-Artes de Lisboa (ESBAL) in 1956, and a Ph.D. degree in morphology of architecture from Faculty of Architecture of Technical University of Lisbon in 1993. He was in Paris in 1964-65 on a scholarship from Centre Scientifique et Technique du Bâtiment (CSTB). He was in Camâra Municipal de Lisboa 1956-62, Ministério do Ultramar 1962-66, Caixa da Previdência 1966-76, Faculty of Architecture of Technical University of Lisbon 1976-97, and invited professor in University of Évora 2004-2005. He realized many projects for public buildings such as kindergartens, elementary schools, institutions for youth, and centres for the elderly. He was a member of Associação dos Arquitectos and Ordem dos Arquitectos 1956-2004. His current interest is in contemporary aesthetics.

Book Review

Branko Mitrović

Serene Greed of the Eye:
Leon Battista Alberti and the Philosophical Foundation of Renaissance Architectural Theory

(Munich, Berlin: Deutscher Kunstverlag, 2005)

Reviewed by Kim Williams

Via Cavour, 8
10123 Turin (Torino) Italy
kwilliams@kimwilliamsbooks.com

keywords: Leon Battista Alberti, Renaissance architectural theory, perspective, Scholastic philosophy, homogeneity of space, Humanism, Aristotelian philosophy, anthropomorphism, Daniele Barbaro

After his previous book, *Learning from Palladio* (reviewed in the *Nexus Network Journal* vol. 7 no. 1), Branko Mitrović has turned his powerful attention to the more difficult Leon Battisti Alberti. *Serene Greed of the Eye* aims at a new understanding of the concepts presented in *De re aedificatoria* through the examination of Alberti's background in Scholastic Aristotelianism. What emerges represents a substantial shift in the usual interpretation of Alberti's treatise of architecture. What also emerges, however, is a polemic on the methods and rationalizations of architecture history itself.

As he did in *Learning from Palladio*, Mitrović opens his examination of the subject with a clear exposition of his methods and aims. First of all, the aim of this book is "the reconstruction of the argumentational framework underlying [Alberti's] theoretical system" (p. 17). This goes beyond the usual examination of terms, words and single concepts that are the usual focus of Alberti studies. If we think of Alberti's treatise as a heavily-laden fruit tree, we might say that while most scholars examine the fruit, Mitrović examines the tree itself, from which the fruit not merely hangs but draws its vitality. The tree in this case is the Aristotelian philosophy of Alberti's time. But somewhat surprisingly, Mitrović argues that while the tree is Aristotelian, the fruit is not necessarily so, and to demonstrate this convincingly he contrasts Alberti's ideas with those of the more orthodox Aristotelian Daniele Barbaro.

Once again Mitrović raises the question of whether architectural theory is concerned with the issue of meaning, or content, which can be verbally described, or the issue of form, which is visual. (The books title obliquely refers to Alberti's preference for the visual. When I queried Prof. Mitrović about the meaning of the

1590-5896/06/020111-4 DOI 10.1007/s00004-006-0023-9

titled, he wrote me that "Alberti says that human eyes are greedy for beauty and Aristotle describes the disinterested enjoyment in things visual. I assume that such greed can be properly called serene. But the original phase / title was "The sovereign jewel of the eye", which is a line in a poem by Geoffrey Scott"; the poem cited is "The Map of Spells".) And as he did with Palladio, Mitrović locates Alberti's interest in the visual. It is only the visual that is capable of entering into the realm of universals; the significance of meaning depends on an interpretation and is thus subjective. The concepts then that Mitrović focuses on to define Alberti's theoretical framework are shape and the elements that define it, that is, lines and angles, and finally, the medium in which shapes are related, space.

One of the problems with which Alberti grappled was that the particular Aristotelian language and philosophical structure he adopted (as the only proper one, Mitrović argues, given Alberti's formal education) was particularly ill-suited for the expression of his ideas about visual universals. In other words, he was trying to express concepts that lay outside the frame of reference of the terms within which he was confining himself. Architects are not so unfamiliar with this situation: consider those, for example, who try to construct a smoothly-curved wall with rectangular-shaped bricks. The curve is approximate, but our minds willingly smooth out the rough edges. Alberti was addressing an audience versed in a philosophical language that was in its nature prejudiced against what he is trying to express, and thus he is limited to expressing it as closely – if not exactly – as he can. He must have hoped that discerning readers would smooth out the rough edges.

Mitrović's hopes to clarify for us the Aristotelian language with which Alberti worked, and at the same time shed light on Alberti's meanings through the careful contrast of the use of conventional terms in *De re aedificatoria* and the use of those same terms by an orthodox Aristotelian such as Daniele Barbaro. By his choice of a conventional philosophical language, Alberti may have been attempting to make himself acceptable to a certain fold or camp of thinkers at the same time that his ideas were distancing him from it.

The contrast between Alberti and Barbaro highlighted by Mitrović tells one side of the story, but this push-pull between language and concepts tempts this reviewer to contrast Alberti with Galileo who, some two hundred years later, enjoined not to express or hold or teach or discuss the notions of the centrality of the universe of the sun or the mobility of the earth, nevertheless does exactly that through the use of a very careful choice of words, combined with irony and satire, in his *Discorsi* (*Dialogues*) – although the cloaks were not forceful enough to protect him from his ultimate painful denial of the ideas that were dearest to him.

Now, obviously language lends itself to both clarity and obscurity, or to revelation or deception. Alberti and Galileo were both addressing readers who were either unused to the ideas being proposed, or downright unbelieving, but while Alberti wished to make his real meanings clear beyond the conventions of his chosen terms, Galileo used the conventions of the terms to cloak his meaning. (It is interesting to hear Alberti voice a complaint against Vitruvius for not being clear enough in his writing: if he wasn't going to write anything that could be understood, he might just have well not written anything at all.)

In addition to issues of linguistics and meaning, however, the contrast of Alberti and Galileo sheds light on the issue of Mitrović's pet peeve, which he calls the "fallacy of collectivism". Like Alberti and Galileo, Mitrović is setting out his argument in the face of prejudices against his basic premise. In his case he is arguing against the body of architectural historians who maintain that all thinkers are conditioned by the times in which they live. One argument of collectivism, and we hear it all the time even when it is not named, is that because we possess knowledge that appertains to one's specific period of time, it is basically impossible to understand a thinker conditioned by only that knowledge that appertains to his own time (that is, we cannot un-know what we know and put ourselves in the shoes of someone in a previous era who did know know what we do). A corollary of collectivism is that it is impossible for a thinker to express ideas that are beyond his time. An example of this is the argument that Alberti couldn't possibly have thought in terms of homogeneity of space because that concept had not yet been formulated in the Quattrocento. To this reviewer, this appears to be a faulty way of thinking on the part of architectural historians, because in the history of the sciences it is assumed that the thinker *has* to think beyond the concepts of this time: the very essence of progress in science is pushing the envelope. On the other hand, there is no doubt that the spirit of the times can condition quite forcefully the way that science is practiced. Alberti and Galileo were both conditioned by the dominant culture of Catholic Aristotelianism. But Alberti didn't risk incurring the displeasure of the Inquisition, as Galileo did; he merely risked being either misunderstood or ignored. This is due to the fact that no matter how unconventional, architectural theory is simply not as dangerous as cosmic theory.

Does *Serene Greed of the Eye* bring us any closer to a definitive understanding of Alberti? Yes and no. Yes, to the extent that understanding the Aristotelian context of Alberti's time is truly helpful in revealing his background and points of departure. For this we can be grateful to Mitrović, because he is one of the very few scholars who have the formal training and background to be able to do this for us. But no, in the sense that Mitrović careful explanation of concepts really cries out for a new translation of *De re aedificatoria*, as Mitrović's comparison with his own translations of given sections of the treatise with that of Rykwert, Tavernor and Leach shows. Until we have access to a full translation in Mitrović's terms, it is really impossible to pronounce a judgement on Alberti's ultimate meaning.

If there is a flaw in Mitrović's argument, I think that it lies in its being occasionally too forced. I am willing for instance to accept the basic argument that Alberti was trying to go beyond his times and its conventions, but descriptions of how the precise wording was determined seem to me to be unfair projections of Mitrović's own thought processes onto Alberti.

But if we concentrate on linguistic problems, or problems of translation, or problems of an author's unknown thought processes, we begin to despair of ever fully understanding what the author is trying to tell us. That brings us dangerously close to the justification of renouncing the possibility of understanding that is the bane of Mitrović's existence.

Although *Serene Greed of the Eye* is difficult, and not without an attitude, it is a welcome addition to Alberti scholarship. If nothing else, it forces the reader to forget the platitudes he has learned about Alberti and think carefully about this extraordinary thinker. Mario Carpo's thoughtful preface provides a very helpful overview.

About the reviewer

Kim Williams is editor-in-chief of the *Nexus Network Journal* and director of the conference series "Nexus: Architecture and Mathematics".

Book Review

George Hersey

Architecture and Geometry in the Age of the Baroque

(Chicago and London: University of Chicago Press, 2000)

Reviewed by James McQuillan

University of Botswana,
Private Bag 0022
Gaborone, Botswana
jasmcq@yahoo.com
MCQUILLAN@mopipi.ub.bw

keywords: George Hersey, Baroque architectural theory, projective geometry, Guarino Guarini, proportion, conic sections, Archimedean solids

With the recent publication of such works as Dalibor Vesely's *Architecture in the Age of Divided Representation: The Question of Creativity in the Shadow of Production* (2004) and Chikara Sasaki's *The Mathematical Thought of Descartes* (2005), we are able to access Renaissance and Baroque mathematics as never before. The earlier appearance of George L. Hersey's contribution to Baroque mathematics unfortunately could not avail of the material of such later contributions, and yet this is a major work that deserves attention. Prof. Hersey has organized his book not as a continuous history, but as a series of 'episodes', interweaving the mathematics, mainly geometrical, with their architectural applications in each episode, and admitting that 'another author might write with equal justification about different episodes' (p. 4). This review will describe his episodes and then discuss what such 'different episodes' might be.

The first chapter, titled 'Introduction', leads on to a presentation of proportion, namely the well-known arithmetical, geometrical and harmonic series. In doing so, Hersey misses out completely when he reaches 'harmonic', as he fails to refer to the philosophical foundations of this topic. This omission reflects on the author's general attitude to his material, the failure to consider ancient and later philosophy to lay bare the primary considerations of such a topic as harmonics, which was for Plato the understanding of growth and order in the world or cosmos. The mimetic importance of harmonics is briefly touched upon when he refers fleetingly to the possibility of cosmic projection in Kepler's *Harmonices mundi*, but the discussion of mimesis, for Hans Johann Gadamer the most important and ancient of all theories of art, is then avoided – 'mimesis' does not appear in the index. Therefore while Hersey provides a wealth of detailed examples of mathematical lore and their applications, the fundamental issues of interpretation and meaning are in the main

1590-5896/06/020115-4 DOI 10.1007/s00004-006-0024-8

absent. With this important observation ever in mind, the remainder of the book will be set forth in order to survey its scope.

The second chapter is 'Frozen Music', discussing Vincenzo Galileo, Kepler and Kircher, and then on to François-Nicolas (not François) Blondel, ending with a modern analysis of Bernini's baldacchino in St. Peter's. There is no reference to music as a Liberal Art, where such relationships were studied for centuries. This chapter, like those that follow, is very well furnished with cuts and coloured diagrams. One application of dissonance from the hand of Piranesi in the 1770s, is 'conscious or unconscious', thus vitiating the inclusion of this base of a candelabra on the architect's tomb (p. 39-40).

'The Light of Unseen Worlds' is the title of chapter three, where he reaches the conclusion that the stacked domes of the Baroque were modeled on telescopes and microscopes. This I find extremely hard to accept, for the simple reason that such instruments were the preserve of a small elite, and therefore meaningless to the general audience of Baroque society. Boullée's cenotaph (not tomb) of Newton appears near the end, and surely this should not be included in a book with Baroque in the title. Hersey never establishes a chronological framework for the Baroque, so such inclusions – this is not the only one – are awkward and puzzling for the average reader. So is Michelangelo, surely a precursor of Baroque, not really a Mannerist? While there is no objection in referring to the great Florentine as Hersey does, such a distinction should be made, so that the reader is warned accordingly.

The fourth chapter, 'Cubices rationes' depends on Kepler's *Harmonices mundi*, and treats of 'tiling' – the fitting together of polygonal and other shapes, flat and solid. He ends up discussing 'Archimedean solids' presumably found in Kepler's text, but we are never certain where such a concept originated, or what exactly are their special attributes beyond the statement that 'the angles of each face . . . are uniform'. But why are they 'Archimedean'?

'Symmetries', the title of the fifth chapter, identifies the modern usage of the term with an expression of Bernini's regarding an altar in Paris in a conversation with the Queen Mother. The following discussion is based on reflective symmetries, and this chapter is perhaps the weakest in the book, as we are never sure if the Baroque believed in such a concept that we now accept, to any general extent. An illustration of Boullée's Palais de Justice is included here but there seems no reference to it in the text.

Conic sections form the basis of the sixth chapter, 'Stretched Circles and Squeezed Spheres'. Its first section is labeled 'The Beauties of Distortion'. This title begs definition – what is beauty? Perhaps a clear consideration of this central and ancient concept would have rescued Prof. Hersey from his aversion to philosophy and provided some manifest conclusions for the reader. The only reference to trammels and drawing machines is found on p. 136, Blondel's compass for drawing an ellipse from his *Cours*. Blondel and many others were fond of such instruments; I will come back to this issue later.

Chapter seven is called 'Projection', a central territory of Renaissance mathematics and art. Here Hersey's episodic arrangements falter, for he discusses Desargues before Renaissance perspective, which would have made a sensible introduction. Another sensible move would have been to discuss the Theorem of Pappus as an introduction to projective geometry, so no doubt the beginner will have great difficulty in negotiating the presentation of Desargues's projection and his perspectival method. It might be added that since the very few such as Pascal and Philippe de la Hire were able to understand Desargues, his impact on actual. building was minimal. Hersey uses Guarino Guarini's method of calculating the area of a circle as an exercise to illustrate projective geometry (pp. 172-173), but it is assuredly not the case that Guarini would have intended this, as his grasp of projective geometry was elusive, extending to the Theorem of Pappus only. To the general reader this exercise will seem perplexing and there is little justification for its inclusion in an otherwise clear treatment of the main material elsewhere. The alleged relationship of regular polygonal circumferences to the outline of gunpowder fortification on pp. 177-178 cannot be at all sustained: the development of Baroque military architecture derived from considerations of angular bastions with connecting curtain walls, and were at first provided to existing cities. That regular layouts were common on fresh sites is undeniable but even so, the precise profiles of solid geometry do not coincide with the particulars of such fortifications as working defenses.

The penultimate chapter is on "Epicycles', deals with circles centered on a greater circle's circumference, and broken symmetry. Hersey struggles with providing some meaning to epicycles, but never refers to the Ptolemaic system that depended on 'saving the appearances' by recourse to such geometries, explicitly in the astronomy of Guarini. His statement, 'After all, rotundas were the earthly homes for the free-flying souls destined for the Heavens' (p. 186), is very amusing but theologically incorrect. The book ends with a chapter on Wright and Corbusier, with the important observation that modern architects cannot think as their forefathers did – computers just approximate what emerges from freehand sketches. Hersey is right of course: his book does illustrate many of the complex activities that accompanied earlier architectural design in so many ways.

So what is missing – what might different episodes address? First, there is the need for a concise treatment of mimesis, how ancient art up until the Enlightenment conceived of beauty as the embodiment of metaphysical and transcendent realities, which in mathematical terms was primarily achieved through harmonics and geometrical imagery of great variety. In dealing with the Baroque a more intensive treatment of light and sound is required – Guarini springs to mind as the great exponent of light at this period. Finally a full treatment of trammels and drawing instruments, such as the mesolabium mentioned by Vitruvius and carried on by Blondel and others including Descartes, to achieve harmonic results, not to mention machines for entasis to full-sized columns, seems important to tie the various procedures together.

About the reviewer

James McQuillan is an architect and theoretician who spent a considerable time in practice spanning 1969 to 1982, which also saw periods in academia, acquiring an MA under Prof. Joseph Rykwert at the University of Essex, and spending a research year in Rome. In that period he was involved in conservation and leisure architecture as well as teaching at his old school of architecture in Dublin, and practising on his own account after 1978. In 1982 he returned to academia full-time, first by teaching in Arabia for four years and travelling in Europe. He completed his doctoral thesis on Guarini in 1991 at Cambridge under Dr. Dalibor Vesely, and has conducted an international career of teaching and research in architecture and the humanities until the present. He is currently B.Arch. Course Leader at the University of Botswana, in Gaborone.

Exhibit Review

L'uomo del Rinascimento.
Leon Battista Alberti e le Arti a Firenze tra Ragione e Bellezza

Palazzo Strozzi, Florence, 11 March – 23 August 2006
http://www.albertiefirenze.it

Reviewed by Marcello Scalzo

Università degli Studi di Firenze
Dipartimento di Progettazione dell'Architettura
50132 Florence ITALY
oycsca@tin.it

keywords: Leon Battista Alberti, Palazzo Strozzi, Renaissance architecture, beauty

Il nocciolo, a cui l'olivo aveva chiesto quando avrebbe dato i suoi frutti, dal momento che fioriva con il freddo, rispose: "Quando sarà tempo"

(The seed, when asked by the olive tree when it would produce fruit, replied: "When the time is right")

L. B. Alberti, *Apologhi Centum*, XXXIV

Ever since the celebrations of Leon Battista Alberti's death in 1972 and, more precisely, since the Mantua exhibit in Palazzo del Te in 1994, studies on Alberti have been increasingly flourishing and successful, both in quantity (number of conferences and workshops), and in quality, thanks to the progress in our knowledge. The founding of the *Fondazione Centro Studi Leon Battista Alberti* in Mantua in 1998 strengthened the common focus of a group of researchers towards the cultural rediscovery of Humanism, understood as the consciousness of the determinant presence of man during the course of history; perhaps inspired by a hopeful desire of its renewal in the modern cultural context, frozen in the search of an identity that combines (ir)rationality and beauty.

This idea appears to be the main theme of the exhibit currently presented at the Palazzo Strozzi in Florence, organized by Cristina Acidini Luchinat, Superintendent of the Opificio delle Pietre Dure of Florence, and Gabriele Morolli, professor of history of architecture at the Architecture Faculty of the University of Florence.

The exhibit itinerary, which flows through an elegantly designed setting, starts with the genealogy and family origins of Alberti, and continues through his literary and scientific formation, his treatises on painting and architecture, his writings on technical and mathematical topics, his literary writings, and his architectural works.

Nexus Network Journal 8 (2006) 119-122
1590-5896/06/020119-3 DOI 10.1007/s00004-006-0025-7

The works are presented from a standpoint that constantly refers to the city of Florence and the Florentine clients, and emphasizes a critical examination of the (ethical) aesthetics of Alberti. There is a web of references and works by other authors derived from – or merely influenced by – Alberti's thought, in order to show how much he infused the artistic vision of what would soon become the Renaissance. Alberti's aesthetic cannot be understood if detached from the moral, political, scientific and pedagogical context upon which he founded his particular classicist culture, but he also understood how, with surprising far-sightedness, to relate to the reality of the late Middle Ages in order to be able to convey it to true Humanism.

La bellezza è l'armonia tra tutte le membra, nell'unità di cui fan parte, fondata sopra una legge precisa, per modo che non si possa aggiungere o togliere o cambiare nulla se non in peggio (Beauty is that reasoned harmony of all the parts within a body so that nothing can be added, taken away or altered, but for the worse) L. B. Alberti, *De re aedificatoria.*

Alberti, literate and erudite, emerges from the Florentine Renaissance in which the main protagonists were magistri, craftsmen, luxury artisans. Albert was far from being a *omo sanza lettere*, as Leonardo used to describe himself.

More than in the scholarly presentation of the Alberti's architecture, the value of this exhibit lies in the release of some new biographical documentation on the Alberti family, a family whose gloomy presence constantly shadowed Leon Battista in his never-ending exile, not so much from Florence as from his own people.

Besides the family papers and the literary writings, the exhibit displays works by artists who were contemporaries of Alberti's, in order to show the complete cultural scene of Florence in the Quattrocento.

The exhibit itinerary – from architecture, paintings, drawings, models, book chapters – shows Alberti as a man often misunderstood in his own epoch, in which he is repeatedly considered as a theoretician rather than as a *magistro*, working master, while at the same time he is revealed to be a person of deep intuition about the reality of the future, based on a knowledge of the past. Alberti is like a seed that blossoms after the dormancy of winter, and especially in his preparation of the ground for the time when fruit would mature.

Sure enough, among the fruits that we may still pick today from the whole Renaissance legacy of the Florentine Quattrocento, both material and hidden within the architectural design process, the exhibit staging could not have been designed without using a proportional system based on the Florentine braccio, as the didactic captions explain to the unprivileged and unlearned visitor, unable to perceive in the exhibit itinerary the underlying harmony of measures and spaces.

The exhibited works are many, all extremely valuable and precious, but not all directly related to the theme of the exhibit and not all worthy such a high-level cultural event. There are some inexactitudes and gross generalizations (not a mention of Laurana, the presumed author of the famous painting *La Città Ideale* !) but of course the average visitor does not notice these. This kind of exhibit is intended for the casual tourist, almost always a foreigner, who will find in the rooms of Palazzo Strozzi a *summa* of the Florentine Renaissance, complemented by effective didactic panels (who reads any more?), many images, a suggestive atmosphere, multimedia movies – spectacular but not always rational – a proud display of professionalism, erudition and technique.

The big exhibits, the big events, the big numbers: these are the cultural abc's to which we are all entitled.

Translated from Italian by Sylvie Duvernoy

About the author

Marcello Scalzo is an architect. He earned his Ph.D. at the University of Florence, discussing a dissertation on the study of the Rotonda by Brunelleschi in Florence, starting from a drawing by Giuliano da Sangallo. He currently teaches techniques of architectural representation and measured surveying at the architecture faculty of the University of Florence. Besides being a specialist in Italian Renaissance architecture, he has also inquired intothe characteristics of rock architecture (Architettura rupestre) of southern Italy. He is the author of numerous papers published in international journals and in the proceedings of international meetings.